DATA REIMAGINED

BUILDING TRUST ONE BYTE AT A TIME

DATA REIMAGINED

JODI DANIELS AND
JUSTIN DANIELS

Legal Disclaimer

The information provided in this book does not, and is not intended to, constitute legal advice; instead, all information, content, and materials available in this book are for general informational purposes only. The readers of this book should contact their attorney to obtain advice with respect to any particular situation. No reader of this book should act or refrain from acting on the basis of information contained in this book without first seeking legal advice from counsel in the relevant jurisdiction. This book is not intended to apply to your particular situation and only your legal counsel should be advising you with respect to your particular circumstances. The views expressed in this book by the authors are those of the individual authors writing in their individual capacities only. All liability with respect to actions taken or not taken based on the contents of the book are hereby expressly disclaimed.

Copyright © 2022 Jodi Daniels and Justin Daniels
All rights reserved.

Data Reimagined
Building Trust One Byte at a Time

ISBN 978-1-5445-3484-8 Hardcover
 978-1-5445-3483-1 Paperback
 978-1-5445-3485-5 Ebook

This book is dedicated to two generations of our family.

To our parents—
for all your support helping us dream big dreams
and pursue them with passion.

And to our kids—
always pursue your dreams with passion
and don't be afraid to take a risk
and put yourself out there.

CONTENTS

INTRODUCTION 1

PART ONE
TRUST

CHAPTER 1
TRUST IS POWERFUL 19

CHAPTER 2
TRUST IS HARD 33

CHAPTER 3
READY OR NOT 53

PART TWO
DATA COLLECTION AND STORAGE

CHAPTER 4
DATA COLLECTION 71

CHAPTER 5
DATA STORAGE. 91

PART THREE
DATA ACCESS AND USE

CHAPTER 6
DATA ACCESS 107

CHAPTER 7
DATA USE. 117

PART FOUR
CYBERSECURITY

CHAPTER 8
OF HACKS AND HACKERS 129

CHAPTER 9
WHEN VULNERABLE IS A BAD THING 143

CHAPTER 10
DEFENSE IN DEPTH 165

CONCLUSION. 189

INTRODUCTION

The great holiday classic, *A Christmas Carol*, begins with this preface: "I have endeavoured in this Ghostly little book, to raise the Ghost of an Idea, which shall not put my readers out of humour with themselves, with each other, with the season, or with me."

With apologies to Dickens, we'll raise a more modern ghost and send it off to visit a more contemporary and kindhearted businessman. We'll call him Bob.

One more update—rather than being whisked away to watch, unheard and unseen, a Christmas past, present, and yet to come, Bob was transported to a parallel dimension where the literary is literal. He'd been sent to see, equally invisibly and inaudibly, the ghosts of data present and past.

In our little story, Bob found he had become a ghostly presence standing with his guide at his own bedside while a parallel version of himself (let's call him Rob) sleeps peacefully next to his wife

Maggie. But Bob and the Ghost of Data Present weren't the only shadowy figures in the room. A man in a trench coat and fedora slouched against the headboard, notebook in hand.

Who the hell is that? Bob wanted to know. *What's he doing?*

The Ghost of Data Present shifted Bob's position to show him the man's notebook. In it, the ghostly spy had logged Rob's heart rate and body temperature in ten-minute increments for the last seven hours. Rob snored, and the data detective added a tick mark to a running tally.

Then Bob noticed another shadowy person standing over his wife's sleeping body. Bob rushed the guy and found himself hovering outside the house. His guide stepped through the wall and pulled him back into the bedroom. "Right," Bob said. "Disembodied. Got it."

Bob resolved to relax. There wasn't anything he could do about the creepy ghost guy watching over the parallel universe Maggie, and he was curious about what his mysterious ghost guide was trying to prove. He closed his eyes and waited.

When his alarm went off, Bob smiled. "I was having the weirdest dream," he mumbled to Maggie.

INTRODUCTION

"Alexa, what's the weather like?" Bob heard Rob say.

Bob opened his eyes and saw he hadn't been dreaming. Or he was dreaming still. Either way, he watched Rob carry out his normal morning routine. Rob asked Alexa a couple of questions, as Bob usually did, and the shadowy detective wrote down the questions and answers. When Rob went into the bathroom, the detective followed and jotted down his weight, reading it over his shoulder from the app on his phone. The number was a bit higher than Bob normally confessed to.

"Nobody's going to see that, right?" he asked, but his guide just shrugged.

In the kitchen, Rob told Alexa to play NPR as Bob did every morning, and he saw the detective make another checkmark. Starting to enjoy himself, he floated through the kitchen table to read the notepad. On the detective's scribbled grid, Bob noticed a gap of several days and recognized it as the long weekend his family had recently taken. "Who sees this?" he asked the ghost. "Anyone who knows I always listen to the news in the morning could tell I was out of town by looking at this. We put lights on timers to keep it from being obvious, but this would be a dead giveaway."

Rob opened the new fridge Bob bought last month, and the detective flipped to a page detailing everything in it. He deducted a few ounces from the weight of the bagged coffee. Rob adjusted the thermostat—Maggie was keen on saving energy and had raised the programmed settings, but the day was already hot, and Rob liked to come into a cool house after his morning jog.

"I once visited a man," his ghostly guide said, "Whose wife had the logins for their thermostat. When they separated, he stayed in the house, and she made things, let's say, 'pretty chilly' for him."

The guide chuckled, but Bob wasn't sure it was funny. He and Maggie were solid, but he could imagine living in a house whose temperature he didn't control—the discomfort and the expense!

Rob went out for his run, the detective carefully noting down the precise time that he unlocked the door and that he didn't lock it behind him. Bob was about to follow but stopped as his daughter appeared at the top of the stairs. "Good morning, Sweetheart!" Bob said. Then, "Wait! Who's this creep?" Because another version of the same data detective stood just behind his little girl. "This is not okay!" Bob turned to his guide. "I'm going to kill that guy if he spies on my kid."

The guide wasn't troubled by Bob's anger. "You signed her up for it."

"No, I—"

INTRODUCTION

"Didn't read the terms of service, did you?"

Bob lunged at the guy and found himself in his office. Rob was at his desk and no detective stood behind him.

"That's better," Bob said.

"Give it a minute," said his guide.

The data detective walked into Bob's office. "How's it going, Bob?"

"He can see me?" Bob asked his guide.

"Going great!" said Rob.

The data detective unslung a bag from his shoulder and opened it over Rob's desk. Hundreds of identical notebooks poured out. They mounded on the desk. They fell onto the floor and Rob's lap. And they kept coming.

"What's he doing?" Bob asked.

"He's delivering the data your company collected overnight."

"We don't spy on people!" Bob was aghast.

"Of course you do," the ghost said.

We'll leave our parable here before Bob gets whisked back to the past, into the offices of Facebook and the lair of hackers because we suspect that nobody picks up a book on data privacy and security for the fun stories. There will be others, but we hope this one has served its purpose—to introduce you to the different and often conflicting ways people understand and experience data.

To Bob-as-consumer, data collection is largely invisible. We believe that if he were more aware of how much personal information is being collected about him—about all of us—it would cause him some concern, but we're not here to data-shame anyone. We get it. In fact, part of what makes our jobs as privacy and security professionals difficult is the amount of finger-pointing and superior sneering our industry inflicts on people who are just trying to keep up with the speed of technology. We understand that many of our peers come off as cranks or paranoid elitists who have no idea what you're doing but you're doing it wrong. We're not like that. We promise.

If we'd followed Bob back to the Years of Data Past, we'd see that before our own fabled "digital age," the information businesses collected about people was more obviously either public or private. Back when mail-order catalogs were the only alternative to shopping at the mall, in giving your shipping address to the phone operator, you knew you were agreeing to have more catalogs mailed to you. (In fact, they used to threaten to *stop* sending them if you went too long between orders.) On the other hand, if you bought something at the mall, you didn't expect your address or phone number to be taken from your check or credit card carbon. Short of bouncing a check, you expected that number would be kept private. Before information was digital, the average person's data was rarely considered valuable enough to sell or steal.

Today, like Bob, most business leaders continue to imagine their own, personal data as an extension of their public identities. And they think of the data they collect from customers as a useful business tool. Some recognize that to companies like Amazon, Google, and Facebook, data is a commodity, and many pay for the use of or access to that resource in the form of advertising. Increasingly, smart business leaders are recognizing another population who has yet another way of imagining data. To hackers and enemy states, data is power—the power to hold business operations or public infrastructure ransom or to influence the outcome of an election.

At the beginning of our Information Age, the tech-savvy founders of companies like Amazon, Google, and Facebook recognized the coming changes and their implications and reconceptualized data as a commodity. They made billions and built empires on that understanding. Alert business leaders today are well-positioned to do the same—to reimagine data again, not as a commodity, but as a relationship. If consumers understand their data as an extension or expression of themselves, how you treat their data is how you treat them.

Like early industrial companies, early information brokers have an exploitative relationship with the resource they mine and sell. Today, consumer groups and government agencies are noticing the data equivalates of smog, acid rain, and superfund sites—spam, identity theft, and ransomware. Individual consumers are likewise starting to seek out companies that have a healthier and more sustainable relationship with their data. This, in turn, is changing how they (and therefore businesses) think about data privacy and security.

To fast-moving internet start-ups hoping to capitalize on everything the digital revolution made possible, privacy and security were impediments or even active antagonists to their business goals. But businesses interested in building the kind of relationships with customers that will carry them into the future recognize that all relationships are built on trust and that data privacy and security are how they demonstrate respect for their customers and build that trust.

Data Reimagined will teach you how to leverage your company's privacy and security practices to strengthen your relationship with customers by becoming (and remaining) worthy of their trust. Of

INTRODUCTION

course, trust isn't easy. People know companies sell their data and start out defensive. Threat actors are actively working to undermine trust, and new technology does so incidentally. Few executives have the time or interest to keep pace with evolving privacy and security demands and find both the procedures and the people they hire to implement them difficult.

But this is good news. If building trust through privacy and security were easy, your competitors would be doing it, and doing what's difficult for your customer's benefit builds more goodwill than doing what's easy. It can also be a competitive advantage in winning B2B contracts. Even more significantly, enhanced privacy and security will soon be mandated for all businesses, and being on the leading-edge positions you to reap benefits that will be diluted as the practices you implement voluntarily are eventually made mandatory.

We'll show you how to build trust with privacy and security in your data collection and storage and how to safeguard that trust in your data use and sharing practices. We'll explain the most common vulnerabilities companies face and outline appropriate protective measures. And we'll do it all with a grounding in practical business concerns and a few more fun stories.

First, though, it's time to kill off Bob's ghostly guide and introduce your very living ones: We're Justin and Jodi Daniels, a husband-and-wife team of data privacy (Jodi) and cybersecurity (Justin) experts who have a combined fifty-five years of business experience working with companies of every size as well as governmental agencies from the federal to the municipal.

Jodi: I'll go first.

Justin: You usually do.

Jodi: I'll introduce you, and then you can introduce me.

Justin: You mean I get the last word?

Jodi: Not even close.

Justin: Okay, just remember what they say about pay-backs.

Justin has both a law degree and an MBA and started out as a business attorney specializing in corporate mergers and acquisitions. As a forward-looking guy, he quickly gravitated toward technology companies helping them deploy everything from fintech to drones. It was his interest in cutting-edge technology and his familiarity with the law that put him in the perfect position to start really paying attention to cybersecurity before it was cool. The more he focused on it, the more he saw that not only was it going to be an important piece of the contracts he'd be working on in the future, it was going to be central to every business transaction. He was fascinated by it and passionate about it, and he started getting a reputation for being ahead of the curve, talking about cybersecurity as a strategic enterprise risk. Companies were flying him to Israel and England to speak on cybersecurity topics not because he knew more about it than the experts, but because he understood (and could explain) it from a business perspective.

In the wake of the SolarWinds and Colonial Pipeline hacks, the rest of the business world started paying more attention to cybersecurity threats. Plenty of law firms now have cybersecurity attorneys, but it's all they know. They're not business attorneys, so while they can manage the privacy and security addendum, they don't understand the practical concerns that businesses have to deal with or contribute insights on the overall business deal. Justin's one of the very few attorneys out there who has deep experience with both pieces so he can look at a transaction and figure out how to balance security with what it takes to get the deal done.

He created a cybersecurity accelerator at his law firm and started Cybercon, an annual conference that brought in guests from around the world. That conference expanded into Atlanta Cyber Week, which hosted events across the city to highlight our cybersecurity ecosystem and succeeded in ways we didn't anticipate. It led to a TEDx talk and to his involvement in even newer technologies. It also generated even more demand for his unique viewpoint which, in turn, led to one of the reasons we're writing this book.

> Justin: Wow, thank you. That was really nice.
>
> Jodi: I'm a nice person.
>
> Justin: You know, I forget sometimes...
>
> Jodi: You forget I'm nice?
>
> Justin: I forget I'm weird.
>
> Jodi: I never forget that.
>
> Justin: No, I mean I forget that my combination of JD and MBA plus future-tech-savvy is unusual.
>
> Jodi: Well it is. And you take all that and combine it with a resident expert in data privacy who's smart and funny and...
>
> Justin: That's my cue, isn't it?

Jodi started out in accounting as an auditor for Deloitte and then went to work in finance at Home Depot, doing a lot of business process work around financial controls. She expanded from finance into strategy and strategic consulting, and then into big media strategy at Cox Enterprises, a big media conglomerate. She then took those skills to Autotrader.com building behaviorally targeted ad networks for them. That was in 2008 before Facebook really got into the game, and before the ad industry started thinking about self-regulation, so she was way ahead of her time. She saw what was coming and got curious about the implications of data privacy both for companies and consumers.

Jodi took that curiosity and created a role for herself where she worked, building a privacy program that expanded until, by the time she left, she had twenty-three brands she was responsible for. She went from there to Bank of America, where she was their expert on digital privacy, working with artificial intelligence. When companies like Google and Facebook started getting into trouble over privacy issues in the EU, Jodi saw it was just a matter of time before the US would start passing laws to regulate it and knew there was going to be a huge need for advice about the intersection of digital marketing, data strategy, and privacy. And since she had experience in data privacy, data strategy, and digital marketing, she set up her own consulting shop to help companies figure out what it all meant to them.

By the time Jodi struck out on her own in privacy, I was also spending almost all my time in cybersecurity. We weren't in exactly the same field, but in closely adjacent ones, and we each found our work informing the other's. It was like having two brains for the price of one, and people caught onto it. We started getting invitations to give joint presentations on privacy and security, tag-teaming each other, and having fun with it. People started calling us the "Jodi and Justin Show," so we went with it! We formalized our combined (and sometimes conflicting) perspectives as a podcast in November 2020 and were speaking at RSA in June of 2021.

> Jodi: RSA is the gold standard of international cybersecurity conferences.
>
> Justin: Right, and then the pandemic hit.
>
> Jodi: And suddenly companies all over the world were scrambling to figure out how to do business with a decentralized workforce.
>
> Justin: And we were out in front talking about the implications of remote work on privacy and security.
>
> Jodi: And since we both have a background in business, we were uniquely positioned to help.
>
> Justin: Which is the other big piece of why we're writing this book.

Data Reimagined isn't a consumer guide or an in-depth technical manual, and it won't provide you with a customized privacy and security plan tailored for your business. It's a high-level grounding in the big issues of security and privacy as they pertain to business that will help company leaders and small business owners understand best practices. We will try very hard not to overwhelm you with technical details or try to talk you through creating your own privacy and security programs, but we'll explain why you need them and equip you with the right question to ask and the background knowledge you'll need to hire the right people.

If you've been handing over personal information to find out what kind of mythological animal you are or using the same password for everything and keeping it on a Post-It note on your monitor, we're not going to scold you. (But please see the tips at the end of Chapter 10 right now!) We understand that you may have avoided learning about data security and privacy because it can be intimidating and complex, and we'll try to do something about the first issue by using illustrative stories and bantering with each other a bit, but technical information is, well, technical. We won't try to teach you more than you need to know, and we promise not to show off or act superior. If you're smart enough to run a company, you're smart enough to learn what we're here to share with you, and we'll make a point of keeping it connected with the practical concerns and constraints of running a business.

In short, we'll walk our talk. Our key assertion is that your relationship with your customers, like all relationships, is based on trust. In the realm of data, trust is created and maintained through privacy and security, and if you have a website or email, at least part of your relationship with your customers exists within that digital landscape. It's exciting territory, full of opportunities to deepen those relationships and extend your reach. We'll show you how. Trust us.

PART ONE

TRUST

CHAPTER 1
TRUST IS POWERFUL

Justin and I (but mostly I) are longtime customers of a national home goods store which, for the purposes of this book, we'll call Box and Keg. We got our first sofa there, stocked our cabinets with their mugs, and shopped there for our first child's crib and changing table. Our youngest is now a pre-tween, and B&K is still advertising nursery furniture to me. I am very much *not* in the market for crib sheets and adorable mobiles, but I don't want to unsubscribe from their mailing list because I see real value in the occasional email they send about the new trends in tween room decoration.

If they asked for the ages of my kids, I'd happily share that information with them because I like and trust the company, but every time they miss the mark, I'm reminded of how impersonal our relationship is. Box and Keg knows when I bought my first nursery item, so it could reasonably deduce I now have a child over the age of five. If it tracked that information and used it thoughtfully, it would stop sending ads for infant gear to me and age up what they showed me along with my kids. This would be an excellent use of data—to deliver better, more appropriately targeted messaging—and it would probably generate an additional sale or two for them.

Recently, another company that we'll call Globe Shop started sending me advertising. In one email, they even asked for exactly the information I'd wished Box and Keg had! But, because I've never shopped there and assumed they simply bought my information from my credit card company, I went out of my way to unsubscribe from their list. Globe Shop hadn't yet earned my trust and was asking for data it didn't deserve. Box and Keg has my trust and isn't asking for data I'd happily share.

Companies that successfully reimagine data as a medium for strengthening their relationship with customers will first earn our trust and then ask the right questions to encourage us to willingly share more of our data with them. These organizations will then safeguard that data, use it for our benefit, and be transparent about that use, deepening their customer relationships by delivering tailored messages that create value for us—and a powerful competitive advantage for them.

Privacy and security experts who don't have an eye on the needs of businesses often make blanket statements condemning data collection, but that's not our argument. We believe in the promise of data and that companies *should* be collecting and using it to personalize their messaging. That said, we're also keenly aware of the dangers to which companies unwittingly expose themselves when they don't handle that data responsibly and with respect for the consumer. When they do, it can deliver additional competitive advantage by keeping companies ahead of (rather than scrambling to catch up with) both increasing legal regulation and customers' changing sensitivity to issues of privacy and security.

THE PROMISE OF DATA

In the "real" world, when we walk into a store, even a local corner coffee shop, we don't expect the barista (who may know our name and usual order) to know how much we make a year, the ages of our children and whether we're in the market for a new car or a Mediterranean vacation. We made a human connection with a human representative of the company.

The online shopping experience is weirdly both less and more personal. Even relatively small e-commerce sites "know" a great deal about us. Technology enables companies to capture a tremendous amount of information, keep it indefinitely, connect it, and share it with hundreds of other companies in the name of connecting us all. Companies use that technology to the fullest, collecting every possible data point whenever and wherever they can, not because they have a clear business need for it, but because they're not sure they'll get another chance. In the rush to gobble up customer data, they forget the customer isn't just data. We're people. Data doesn't shop for swimsuits and sunscreen. People do.

Justin: I'm hearing a theme here.

Jodi: I hear Italy is nice.

Having information about your customers isn't the same thing as having a relationship with them. In fact, it can often have the opposite result. People are increasingly alienated and, to introduce a technical term—creeped out—by how much Facebook and Google "know" about us. If I've spent an hour online researching sunny islands, an unsolicited email from Air Malta doesn't make the airline feel like my friend. It's assuming a level of intimacy it hasn't earned. In googling for information, I'm doing the equivalent of looking at the work a local artist has hanging on the wall of my coffee shop. I do not expect him to show up at my house with a painting.

We believe consumers are yearning for personalized interaction with companies they can believe actually care about them. This is the promise of data that's been appropriately used and respectfully handled. It can give your company the ability to provide unique, individually-tailored experiences that demonstrate your trustworthiness to your customers in the respectful and transparent way you collect their data with their permission, use it to personalize your interactions with them, and protect it from misuse by others.

To users, data is personal. Gradually sharing personal information with one another is how human beings build relationships. Intimacy is a function of disclosure. We open up to people as we feel closer to them and close ourselves off when trust breaks down. The inverse is also true. We feel closer when we share information about ourselves and erode trust when we start to withhold information. Companies that collect consumer data without permission are effectively stealing the currency of intimacy, creating resentment rather than relationships.

THE DANGER OF DATA

Companies aren't wrong about the role data can play in building online relationships. In fact, what they often need is more data to do so effectively. The issue isn't the amount, but the way in which it's collected, stored, shared, aggregated, and used. When users trust companies the way I trust Box and Keg, they'll willingly share additional data. But trust must come first. Building privacy and security into the technology, not as an afterthought, but as an integral part of its design, is the only way to create that level of trust.

Companies invest hundreds of thousands, if not millions, of dollars in cultivating and growing their brand to foster goodwill in their customers, oblivious to the resentment they may be feeding just as aggressively. Worse, many risk the outright rage directed at companies when their lax security exposes their customers' data to hackers and thieves. Failures to treat people's data with respect can destroy in an afternoon all the goodwill and brand equity a company has spent years building.

People want to do business with companies they trust and feel good about. We want to like the companies we buy from. People love to buy from companies like Toms and Bombas because, for every pair of shoes or socks you buy, the company gives a pair away to people in need. In other words, people buy from them because they like the company. Failing to treat your customers' data with respect is a great way to make them not like you.

It's also beginning to cost companies business opportunities. People in B2B businesses think that because they don't sell to consumers, privacy issues aren't relevant to them, but they are. I've seen deals not get closed because the company on the other side of the negotiation was concerned with protecting its users. As an example, in a common scenario, a company is shopping for a survey vendor. Being interested in building trust with customers and knowing that some survey companies aggregate survey data from their client companies and sell it, they want to know that won't happen. Those companies who can guarantee that have a competitive advantage. The same is true for security where companies are increasingly asking more probing questions and for audits and certifications of vendor companies.

B2C companies, on the other hand, are beginning to feel some pressure from consumers. A 2020 Pew survey showed that 52 percent of people would not buy a product or service over privacy and security concerns, and Cisco reports that 97 percent of companies who invested in robust privacy policies credited them for at least one tangible benefit.[1] Additionally, more companies are looking at the ethical questions raised by data collection, use, and storage as part of their ESG (environmental, social, governance) initiatives. They're recognizing, on a purely pragmatic level, that how they handle customers' data says something about their ethical standards and who they are as a brand—not too surprising considering 87 percent of consumers think data privacy is a human right.[2]

---

[1] Andrew Perrin, "Half of Americans Have Decided Not to Use a Product or Service Because of Privacy Concerns," Pew Research Center, April 14, 2020, https://www.pewresearch.org/fact-tank/2020/04/14/half-of-americans-have-decided-not-to-use-a-product-or-service-because-of-privacy-concerns; Cisco and Beacon Group, *Privacy Gains: Business Benefits of Privacy Investment* (2019), https://www.cisco.com/c/dam/en_us/about/doing_business/trust-center/docs/privacy-gains-business-benefits-of-privacy-investment-white-paper.pdf.

[2] Rande Price, "Consumers Believe That Data Privacy Is a Human Right," *Digital Context Next* (blog), August, 17, 2020, https://digitalcontentnext.org/blog/2020/08/17/consumers-believe-that-data-privacy-is-a-human-right.

For most companies, these concerns aren't yet pressing enough to drive behavior change. Of the few who are paying attention, some are starting to recognize exciting opportunities.

ANOTHER COMPETITIVE ADVANTAGE

We have a friend who is a few years older than we are and has happy memories of family road trips taken in an old, wood-paneled station wagon. The car's "way back" had facing seats, and she and her sister would alternately play on the floor and hang the top half of their bodies out of the roll-down rear window. She's not sure the car even had seatbelts.

Today, when she gets in her designed-for-safety, airbag-equipped, crumple-zoned car, she pulls on her over-the-shoulder-and-across-the-hips seatbelt without consciously processing it. This is the level of seismic mindset shift that's headed our way—and for much the same reasons.

While a few car companies were advertising safety features before Ralph Nadar's *Unsafe at Any Speed*, most weren't terribly interested in how their vehicles performed in a crash. They marketed their cars' shiny new features, performance, and speed. It was only after Senate hearings prompted by Nadar's book that the federal

Department of Transportation was formed, resulting in seatbelts becoming mandatory equipment in 1968.

When Mothers Against Drunk Driving started educating the public about their life-saving power, more people started to actually use them, but the first state law requiring drivers and front-seat passengers to wear them wasn't passed until 1984. By the end of the 1990s, over 60 percent of states had passed seatbelt laws. Today, New Hampshire is the only state that doesn't require them for drivers over eighteen. The laws vary widely, state by state, but usage was estimated to be an impressive 90.3 percent in 2020.

Today's internet is where the US auto industry was in the early sixties, and data privacy and cybersecurity are the seatbelts most companies aren't that interested in. Businesses today as then (as always) are profits-driven, and there's not an obvious bottom-line business case for safety measures until they're enforced industry-wide. This is an enormous opportunity.

When in 1990, with European car companies' sales in steep decline, Volvo went the opposite direction. The company credited its long-standing (and at the time, somewhat aberrant) practice of advertising its cars' safety features. Volvo was associated in consumers' minds with safety before they started prioritizing it and while Oldsmobile was still actively and vocally fighting against regulatory legislation.

In this little parable, Apple is Volvo and Facebook is Oldsmobile. Facebook has consistently and repeatedly chosen to prioritize advertisers and its own algorithm over its users' privacy. Apple, on the other hand, has been garnering goodwill, recognizing the very real opportunity to engage their customers around something people are starting to care more about. Many other companies, especially small ones, have been practicing an "Ignore it and hope it goes away" policy, but this is changing.

The first US data breach law was enacted by California in 2002 and went into effect the next year. Today, there are fifty-two such laws (one for each state and two US territories). In 2004, California again led the way with the California Online Privacy Protection Act (CALOPPA), which required websites to carry privacy notices, and a few other states followed suit. In the following years, there has been a flurry of new laws (which we'll discuss in more detail a little later).

In 2021, in the wake of the Colonial Pipeline ransomware attack, the federal government started getting serious about cybersecurity. Companies are already required to deal with all existing regulations, and more are on the way. The lawless "Wild West" days of doing business on the internet are coming to an end, and companies that get ahead of the new laws will enjoy a significant competitive advantage.

To do that, they'll need to start attending more carefully to how they collect and use data (privacy) and how they prevent unauthorized access to it (security).

Data Privacy and Security Differentiated

While we're sure there's an analogy to be made to our marriage, the easiest way to explain the relationship between data privacy and security is to think about it like your house. If you have an alarm system, three locks on every door, and bars on the windows, your house is probably fairly secure. You've made it as hard as possible for bad people to break in and steal your stuff or hurt your family. But if anyone can peer in from the street because you leave the curtains open, your house isn't private.

And maybe that's okay with you. It's just the living room, right? You probably don't have a street-facing plate glass window in your bathroom. If you do, I bet it's high in the wall or kept curtained. Likewise, you probably don't mind having your first and last name and photo on LinkedIn, but you probably strip your street address and phone number out of your online resume and don't include your height, weight, and last blood pressure reading.

Your doctor's office, on the other hand, does have that information about you on its computer, and if it's properly set up, the medical staff has access to it, but the receptionists do not. They're probably lovely people; they'd never break into your house, but they're not authorized to look at your medical information, and making sure they can't inadvertently access it is a security concern. To return to our house analogy, if you host a block party and move your prescription meds to the upstairs bathroom, you're hoping for privacy. If you lock them in a drawer to make sure nobody sees (or steals) them, you've secured them.

The distinction can get subtle. If someone uses your upstairs bathroom during the party and reads all the pill bottle labels, it's a breach of privacy. If they steal a pill or two—from the medicine cabinet or the locked drawer—you've had a security breach. To use a more business-relevant example, if you walk into REI and buy a ski jacket with your credit card and then get flooded with brochures from Telluride and Breckenridge because REI connected your credit card data with your email address and sold that combination of information (that this person at that email address bought this coat), REI has violated your privacy. Their security people may well have done their jobs if the databases that hold all your information are adequately protected and don't get hacked, but the company has still jeopardized consumer trust. Although what they've done isn't necessarily illegal (depending on their initial disclosure), it's a violation of our expectations.

CHAPTER SUMMARY

Companies that can reimagine data as the new medium of trust decide to proactively care about privacy and security, recognizing the promise it holds out to build trust and demonstrate respect for their customers. They make it part of their sales pitch to other businesses and consumers alike, and get ahead of changing legislation, avoiding the costs in time and money of trying to retrofit their operations.

Now, more than ever before, business relationships (like all relationships) are built on trust. Trust has never been easy, but the ubiquity of data and rapidly changing technology—the very things that make it more important—also make it more difficult. Here again, the answer is to reimagine data.

CHAPTER 2
TRUST IS HARD

It's tempting to draw parallels between good data privacy and cybersecurity practices and other, older, or more familiar kinds of preventative action. Nations prepare for the possibility of war by maintaining a standing army and forging political alliances. As individuals, we prepare for the possibility of a car accident by wearing a seatbelt and for the chance of heart attack by maintaining a healthy diet and exercising.

Jodi: Oh, *we do, do we?*

Justin: Okay, you do. I try to. But you see what I did there? I saved the best analogy for last.

Jodi: Right. Because even after people have been warned by their doctor that only lifestyle changes will prevent a second heart attack, only a small percentage of people are able to stick to their prescribed diet and exercise plans.

PART ONE: TRUST

Shakespeare: If to do were as easy as to know what were good to do, chapels had been churches and poor men's cottages princes' palaces.

Jodi: What he said.

Justin: Who let him in?

Shakespeare exits (pursued by a bear).

Online privacy and security experts have spent years advocating for the equivalent of daily walks and leafy greens with only marginally better results and against greater resistance. Even the most basic privacy and security measures introduce some degree of impedance. Many are in direct conflict with the premium business and industry put on efficiency. Most require some degree of financial investment. To the average CEO, the dangers seem distant and abstract while preventative measures are immediately inconvenient, expensive, and confusing.

We get it. Building trust has never been easy. In this chapter, we'll take a closer look at exactly what makes it particularly difficult for today's business leaders: the perception (and occasional reality) that what's sexy isn't safe and what's safe isn't sexy, the human mind's limited ability to accurately gauge risks and take appropriate preventative measures, and the nature of technology itself.

THE SEXY FUTURE

On a morning in the future, you (or one of your children) will wake up to big news. While you slept, the smart contract* for the purchase of your dream home has been executed and ownership transferred. Your biometrics will now activate the smart locks. You hop in your car, push the button that puts it into fully autonomous mode, and take your first meeting of the day by video.

> *Smart contracts are block-chain-enabled programs that automatically execute an agreement so that all parties can be certain of the outcome once all the specified pre-conditions have been met.

Your new house is a hundred miles away from the chaos of the city, so you head to the drone port where a discreet camera captures your photo and runs it through facial recognition software so there's no wait to go through security. You jump in a pilotless quadcopter, slip into a VR headset and try out different interior designs for your new living room in an exact virtual duplicate populated with furniture models and paint colors from two dozen retail stores that can have anything you order to your new address overnight.

PART ONE: TRUST

Sounds great, right? Sign me up. Except…

On a morning in the future, you (or one of your children) will wake up to big news. Someone hacked into the quadcopter your spouse was planning to take and crashed it. Reeling, you stagger out of bed and hear your front door's smart lock whir. The police have used the GPS data from your car to locate you and the biometrics from your driver's license to unlock your house. Facial recognition software identified you at a peaceful political rally* and the authorities have a few questions for you.

> *In San Diego, smart technology that was designed to regulate streetlights was used to identify and target people who'd attended Black Lives Matter protests. China and other authoritarian states find the level of surveillance technology makes possible very sexy indeed.

Technology is cool. It's forward-looking and optimistic and promises to save us time and make our lives frictionlessly convenient and efficient. Its very brightness can blind us to its dangers. And it doesn't help that the people raising concerns about this exciting future often wear figurative if not literal nerd glasses—very not sexy.

Mutual Assured Incomprehensibility

The C-suites and the IT departments of most companies speak different languages, have different (and sometimes mutually exclusive) relationships with technology, and often have little patience with each other. It's not quite a cold war, but we've both seen executives whose definition of success for their IT department hinges on their own ability to send and receive email. What they really want is for IT to "handle all that," and they don't like talking to techies.

These are smart, often MBA-holding folks but they find what the IT department does both hard to understand and worse—boring.

Like their C-suite counterparts, people who work in the upper echelons of IT are also usually highly intelligent and expertly trained. But they often struggle to explain what they do at an appropriate level of detail or to contextualize it for upper management. It's critically important that these people be highly detail-oriented, but their CEO doesn't need to know what a firewall is or how one works. They need to know why it's important and the way it relates to bigger issues and business goals.

It's extraordinarily rare for anyone outside a company's IT department to truly understand how its network works or how data flows through the organization, and that's assuming someone inside IT knows, which isn't always the case.

> Justin: I once did a cyber proctology exam because—
>
> Jodi: Don't say it.
>
> Justin: C'mon, it's funny.
>
> Jodi: It's not.
>
> Justin: Because I'm Cyber Ass Man.
>
> Jodi: Nope.
>
> Justin: I still want that license plate.
>
> Jodi: Just tell the story.

I once did a forensic workup on a venture-backed company and discovered that one of the employees had been using idle computational power to mine bitcoin, and nobody had even noticed.

Privacy laws are usually the bailiwick of the legal team, and they face the same language barrier in trying to communicate with the marketing and product teams. For example, there are specific legal definitions of what's meant by "consent," on which the marketing team must execute despite their brand language and to which the product teams are bound. But much of what erodes trust is completely legal, so privacy experts often need to make values-based decisions, to both interpret and extend the law in particular circumstances.

Anti-Fast

Human beings continuously and restlessly develop and adopt new technologies to make our lives faster and easier. Efficiency is the goal of innovation, but the measures needed to protect users' privacy and data often come at the cost of efficiency and ease of use. Two-factor authentication could prevent 80-90 percent of cyber attacks,[3] but many companies feel the few additional seconds it requires is too much of an inconvenience to inflict on their customers. And they're not wrong. Many users do, in fact, complain about needing to type in that four-digit code.

For the CEO focused on getting a new product to market, privacy and security considerations feel like an opportunity to introduce additional aggravation and delays.

Reserving the assessment of privacy and security concerns until the end stages of product development heightens this sense. We'll talk about how to bake them into the DNA of both companies and

---

[3] John B. Roberts, "Save Money While Improving Security: The Value of Multi-Factor Authentication," *Coalition Blog*, Coalition, September 10, 2019, https://www.coalitioninc.com/blog/save-money-while-improving-security-the#!.

products in Chapter 10, but until organizations start to do just that, they'll likely continue developing new products in conjunction with their marketing and product teams, turning them over to engineering and then, close to the finish line, taking a quick detour to "run it past legal." When the lawyers throw the red flag, privacy and security look like what tripped you up. They aren't. Not thinking about them from the outset is the problem, but until that changes, privacy and security are going to feel like something that just slows you down.

Because technology evolves rapidly, privacy and security need to keep pace with innovation. There cannot be a "one and done" solution, nor is it enough to simply put an off-the-shelf plan in place. A company's location and the kind of data it collects change how it's legally required to keep its users' data private. An e-commerce and professional services company will have different security priorities. In Chapter 10, we'll talk you through the outlines of how you'd start to tailor a privacy and security plan to your company's unique needs, but for now, it's enough to understand how the perceived slowness and unsexiness of privacy and security make trust hard while even well-intended legislation makes it even harder.

It's Complicated

Considering that the world wide web is, in fact, worldwide, it regularly surprises clients that it matters where they're located, but it does. Different privacy regulations are in scope for different companies in different places. A small chiropractic office whose clients all live in the same state where the office is located will have different needs than an e-commerce company that doesn't ship internationally which will, in turn, differ in important ways from one that does. The e-commerce sites, on the other hand, probably won't need to concern themselves with the special class of regulations that govern the handling of health data, some of which may or may not require them to deal with HIPPA laws.

RISK ASSESSMENT

As an American, you're more likely to die by drowning in the tub or from being buried alive than from foreign and domestic terrorism combined,[4] yet very few restrictions are placed on the purchase of shovels or rubber duckies. More than twenty years after Richard Reid failed to set off explosives hidden in his sneakers, we're still taking our shoes off at the airport. This, in part, is due to how easy it is to imagine a terrorist attack and how hard it is to see a killer

---

[4] In the US, 403 people drown every year while taking a bath and 29 are killed by being buried alive, while right-wing extremists and Jihadist-inspired attacks by foreign and American-born terrorists (2.6 and 2.4 respectively) kill five people each year. Emmanuelle Saliba, "You're More Likely to Die Choking Than Be Killed by Foreign Terrorists, Data Show," NBC News, February 1, 2017, https://www.nbcnews.com/news/us-news/you-re-more-likely-die-choking-be-killed-foreign-terrorists-n715141.

tub in your mind's eye. Perhaps the only threat harder to visualize is that which data privacy and cybersecurity strive to prevent.

For the first decade of the internet as a business tool, most executives have had a vague sense that their companies were exposed to cybersecurity and privacy risks but weren't sure what those risks were or what made them so risky. When it comes to ransomware, 2020 seems quaint when ransomware increased by 92.7 percent in 2021. The dramatic, steady increase in ransomware attacks[5] and the high-profile Colonial Pipeline hack have combined to make privacy and security harder to ignore, but most leaders still deal with tremendous amounts of internal resistance to taking appropriate protections.

The hundreds of hours of downtime and recovery and thousands or millions of dollars of damage a hacker could force on companies and consumers alike aren't compelling enough to get us into action. They feel as far away as retirement and as unlikely as death by drowning in the tub.

They aren't.

---

[5] "Ransomware Attacks Nearly Doubled in 2021," *Security*, February 28, 2022, https://www.securitymagazine.com/articles/97166-ransomware-attacks-nearly-doubled-in-2021.

Today, there are enemy states, hackers, and malicious programmers actively working to undermine the trust consumers have in the companies they do business with online. Chapter 8 will explore the human threats to data privacy and security in more detail, but these bad actors and their increasing proficiency and ubiquity are certainly part of what makes trust so difficult to build. Worse, hackers and criminals aren't the only people complicating your trust-building work. Your privacy and security are only as good as that of the least-protected company with whom you have a data relationship.

ANTI-TRUST TECH

Technology almost by definition is inhuman and impersonal, and that alone makes any relationship mediated by technology feel less trusting. Add to this inherent distrust the stories that keep breaking about how mercenary many companies are with the data they collect, and any initial innocent trust people may have once had evaporates.* Facebook is the perfect example.

PART ONE: TRUST

> *A Pew poll showed that 69 percent of consumers believe most companies will use their personal information in a way that they are not comfortable with.[6]

Facebook (almost ironically) started with the stated goal of connecting people to one another. Because initially, an ".edu" extension was required to join, it suggested a safe, collegial environment where first Harvard and then other East Coast college students could locate and befriend (or, in its new verb form, simply "friend") their classmates.

At the time, nobody imagined how great a reach and deep an impact Facebook in particular and social media, in general, would have in fewer than twenty years. It's now most people's primary news source and had an outsized influence on the outcome of presidential elections. But people joined the platform to connect with people. That connection meant they also shared a stunning amount of personal data. And Facebook sold it.

---

[6] Brooke Auxier et al., "Americans and Privacy: Concerned, Confused and Feeling Lack of Control Over Their Personal Information," Pew Research Center, November 15, 2019, https://www.pewresearch.org/internet/2019/11/15/americans-and-privacy-concerned-confused-and-feeling-lack-of-control-over-their-personal-information.

As it became increasingly obvious that the company prioritized profits over absolutely everything else, many of the people who used the site did so with growing distaste. Some dialed back their usage. Others quit—or tried to.

> Justin (singing): Welcome to the Hotel Facebooka-fornia
>
> Jodi: It is a little like that, isn't it?
>
> Justin (singing): You can check out anytime you like, but you can never leave.

If you have a Facebook account, even if you rarely visit the site, you're probably still logged in. And being logged into Facebook allows the company to track you, your shopping habits, interests, and location across the web and in the world. And they're not simply passively watching you and monetizing what they learn. They're also using what they know about you to do everything they can to keep you coming back.

Facebook has invested heavily in an algorithm that's explicitly designed to addict its users. It deliberately inflames political anger, and they aren't the only ones. Companies use our data to game our biology to create things that upset us to increase our engagement with their app, and we know it. It also creates distrust.

But it goes beyond simply manipulating us by using our data to build an algorithm that pushes content based on our likes and choices. These companies track our interactions with the system to build deep and detailed profiles which it then sells to advertisers so they can more effectively market to (read "manipulate") us too. And they haven't asked our permission to do any of this.

In our not-so-sexy future story, the traffic cams that led to your arrest didn't ask your permission to film you or to run your image through their facial recognition software. In the present, Google Maps never asked me if it was okay to take pictures of my house to use on its Street View and my neighbor's Nest camera doesn't need my okay to film me, even if the Joneses later decide to post that clip on Next Door. Outside our neighborhood, the city placed a camera with digital license plate technology to identify criminals driving their cars. It also captures every time we enter our neighborhood which the city could use to identify patterns of when residents come and go.

And it's not just the volume of data that's being collected that invades our privacy and threatens our security. It's the kind of data and the ways in which it's being collected and combined with other data from other sources and subsequently used.

For a shocking number of companies, the data they collect about us is the most valuable asset they have. And this isn't limited to online businesses. Many retail organizations will happily give you a dollar off the occasional item in exchange for your personal data which they collect and update via the loyalty card that entitles you to your discount. This information connects what you bought today with a record of all your previous trips going back years and with your email, phone number, and street address.*

> *A study from Princeton University found that more than 60 percent of the top 1,000 sites on the web share information with third parties, many of which "fingerprint" visitors and sell the resulting data. They also found that 96.5 percent of websites have access to digital fingerprints even if they don't use it.[7]

In one particularly egregious and well-publicized demonstration of this kind of technology's power, Target correlated a seventeen-year-old girl's purchases with those frequently made by other women in similar circumstances and knew she was pregnant before

---

[7] Steven Englehardt and Arvind Narayanan, "Online Tracking: A 1-Million-Site Measurement and Analysis" (paper presented at the Association for Computing Machinery Conference on Computer and Communications Security [ACM CCS], October 2016), https://www.cs.princeton.edu/~arvindn/publications/OpenWPM _1_million_site_tracking_measurement.pdf.

her parents did. They found out when congratulatory ads for diapers showed up in their mailbox. While what they did was entirely legal, had Target's marketing team considered it from within a trust-building paradigm, they might have opted to create different limits for the application of their predictive powers for health-related purchases.[8]

Several retail chains deploy a technology that tracks how many people are in any given store at a certain time. It can connect that data to other information and identify you. If you happen not to have turned off your phone when you walked in (and who does that?), the store can serve a coupon directly to your phone.

There's a strong case for paranoia here, but we promised not to go there. Even without tipping into it, even the least informed people are feeling increasingly suspicious and distrustful of businesses and technology.

Since all relationships are built on trust, how can you hope to build relationships with your customers in that environment? Or do the

---

[8] Kashmir Hill, "How Target Figured Out a Teen Girl Was Pregnant Before Her Father Did," *Forbes*, February 16, 2012, https://www.forbes.com/sites/kashmirhill /2012/02/16/how-target-figured-out-a-teen-girl-was-pregnant-before-her-father -did/?sh=5e63b3e76668.

benefits of technology inevitably come at the cost of our trust in each other, in our government, and in our businesses?

We believe the only way to build trust in this climate is to strike a balance between the sexy promise of technology and the no less sexy promise of respect for one another's data. Trust is powerful, but it isn't easy. This is good news. Precisely because it's hard, demonstrating such respect for your customers' data is an excellent way to signal that you care about them.

Costly Signaling

Apple's scrupulous attention to every detail of its design—the nearly button-less interface, intuitive software, sleek hardware, and elegant packaging—is obviously labor-intensive. People can see the thought and care that went into it and they line up in droves to pay more for it than a comparable and readily available PC.

This is the same principle behind baking homemade cookies for a friend or knitting them a hat. The investment of time and effort creates value and conveys meaning that a package of Fig Newtons or a Madewell beanie doesn't.

In trust-building, this means baking (or knitting) measures to protect your customers' privacy and security into the DNA of your products. It means you invite someone from your privacy, security, and (audible gasp) legal teams to the earliest design meetings to help identify the kinds of data you'll be collecting and to surface where you'll need to ask for consent or include a pop-up notification. Data for which there isn't a business reason won't be collected, while some will be anonymized or not retained, and from the earliest design phase, you'll be thinking about what your privacy notice must include. You'll identify the most critical business processes and data to ensure that they're protected and backed up, implement multi-factor authentication from the outset, and establish a robust security incident response plan.

CHAPTER SUMMARY

If human beings were completely rational, we'd **all have slender bodies and fat retirement accounts**, but most of us aren't and don't. We understand the risks of aging, weight gain, being out of shape, and an underfunded retirement, but the future seems far away. We know we should take actions today to create a healthier, happier tomorrow, but we consistently choose a marginally easier today over a much better tomorrow.

Trust is hard because the people with the power to institute sound privacy and security policies and the people with the expertise to enact those policies have long-standing and entrenched difficulties communicating. Additionally, many of the things that increase trust also decrease speed and efficiency. On their own, these two factors would make it tough to use technology to increase the trust that it erodes. Yet the challenge is made even harder because technology also empowers criminals who actively work against trust. Happily, the very fact of its difficulty makes treating customers' data with respect a powerful demonstration of your company's commitment to them.

CHAPTER 3
READY OR NOT

I (Jodi) was a huge fan of the Jetsons as a child so I was thrilled when, a few years ago, some very clever people realized that between the recent increase in computing power and the availability of more and cheaper storage, the world was finally ready for the kind of videophone I'd seen in the cartoon. Targeted at businesses first and billed as "video conferencing," Zoom's initial pitch wasn't as a replacement for in-person meetings. Instead, the company sold itself as easy to use with an intuitive interface that made it simple to add people to a call, share screens, or otherwise bring a visual element to the then ubiquitous conference call.

Justin: Jodi was an earlier adopter.

Jodi: I was doing a lot of consulting work from home in 2019.

Justin: I have a nice stock portfolio based on my wife's working habits. She's always a year or two ahead of things.

The company was still new and just starting to grow when the pandemic hit in 2020. Suddenly no one was going anywhere. Travel for in-person out-of-town meetings was suddenly impossible and even face-to-face interaction within local teams slowly came to a stop. Companies scrambled to find a way to continue doing business, and individuals who were introduced to Zoom at work were soon Zooming with friends and relatives. (And you know you've arrived as a company when your name becomes a verb.)

Zoom was *the* success story of the early pandemic which had several important implications: shares we'd bought at under one hundred dollars skyrocketed (which was great), hackers started paying attention (which wasn't), and news reporters got curious about it.

Because Zoom, like most tech companies, worships the twin silicon gods of speed and user adoption, it hadn't bothered to build privacy and security into its design. Without safeguards or passwords, hackers found it laughably easy to break into meetings, crash conferences, and show up uninvited in classrooms. Because most of the platform's users were equally devoted followers of the cult of convenience, few had paused to wonder if they needed permission to record a call or what Zoom was doing with the information it collected.

What might have happened if Zoom had prioritized its relationship with its customers over market share? If it had invested in building trust instead of convenience? If it hadn't shared private user information with Facebook and waited until it was sued to institute basic privacy and security features? Sure, it wouldn't have eclipsed front-line healthcare workers or the scientists working on a vaccine, but it could have been heroic. How much trust and loyalty,* even affection, would people feel toward the company that protected their privacy *and* created a secure space for virtual happy hour? It might even have saved them the $85 million they paid to settle a class-action lawsuit alleging they violated California and Federal law.

> *Turns out, the answer is probably at least 84 percent.[9]

US PRIVACY AND SECURITY LAW

The US has taken a sectoral approach to the regulation of data. There are federal laws for certain sectors (HIPAA for health, GLBA for financial institutions, CAN-SPAM for email, TCPA for texting & auto-calls, etc.) while individual states are rapidly creating a patchwork of sometimes conflicting regulations leaving companies

---

[9] Joe McKendrick, "Customers Are Open to Artificial Intelligence Boosting Their Experience, Survey Shows," *Forbes*, June 21, 2019, https://www.forbes.com/sites/joemckendrick/2019/06/21/customers-are-open-to-artificial-intelligence-boosting-their-experience-survey-shows?sh=7f3fd36e5e99.

to understand the differences between them and what to do about those consumers who aren't covered by any existing legislation. To better understand where we're headed, it's useful to take a look at the short but colorful history of where we've been.

In California, almost anything can be placed on a ballot if enough people sign a petition. So, on the heels of the Cambridge Analytical scandal, when the wealthy Alastair MacTaggart got irritated with the liberties he felt Google and other big tech companies were taking by collecting data and profiling people, he set out to get a privacy law enacted. He created a petition, used his wealth and influence to fill it with the required signatures, and got his proposed new law on the ballot.

Of course, this didn't thrill many of the major tech companies that call California home, and they pressured the legislature to pass the California Consumer Privacy Act (CCPA) in June of 2018 to prevent MacTaggart's ballot initiative. Undeterred, and using the same strategy, MacTaggart got the new, more stringent California Privacy Rights Act (CPRA) on the ballot. Considered the first comprehensive US state privacy law, it passed in November 2020 and will go into effect in January of 2023.

Subsequently, other states looked to California's leadership in this arena. While many simply copied California's legislation word for word, some were more thoughtful. Virginia, for example, saw value in California's approach but found it overly focused on the individual. It took parts of the General Data Protection Regulation (GDPR) and pieces of California's new laws, gave them a business spin to match their goals, and passed the Virginia Consumer Data Protection Act (VCDPA) in March of 2021. As of early 2022, twenty-five-plus states have introduced some sort of privacy legislation. While they haven't all passed,[10] the fact that over a quarter of states has considered such laws speaks to the importance of privacy and hints at the future.

MEANWHILE, ACROSS THE POND...

The US and EU have starkly different philosophies about privacy based on their individual histories. During World War II, the privacy of European citizens was grossly violated by the organizations like the Gestapo, and in Russia, the KGB. These secret-service-like groups spied on people and collected private information on them. Consequently, Europe, especially Germany, takes individual privacy rights extremely seriously, and its data privacy laws reflect

---

[10] As of mid-2022, CA, CO, VA, UT, and CT have all passed (subtly different) data privacy laws. "State Laws Related to Digital Privacy," National Conference of State Legislatures, June 7, 2022, https://www.ncsl.org/research/telecommunications-and-information-technology/state-laws-related-to-internet-privacy.aspx/.

PART ONE: TRUST

this. They're strongly individual-focused with business taking a secondary seat.

In contrast, the United States' ranking order for privacy is business first, government second, and *then* the individual. While the EU views privacy as a fundamental right, Americans are much more comfortable with government intervention, including backdoors* and wiretaps. Our federal government has agreements in place with many tech companies (e.g., Facebook, Apple, AT&T, etc.) to gain access to their customers' data. While it's for a good reason—the FBI needs to track terrorists and whatnot—the EU takes heavy issue with America's policies and remains skeptical of our practices concerning digital privacy.

> *What's a Backdoor and Why Does It Matter?
>
> Imagine you designed an impenetrable steel fortress, but you wanted to leave a way in just in case you went out and forgot your keys—you might hide an entrance around the back, right? In some out-of-the-way spot where people wouldn't think to look. In technology, a backdoor is essentially the same thing—a hidden way to access something that is otherwise (and seems) inaccessible.
>
> For example, if the government found that a certain

> person was a threat to national security and needed access to their information, they could demand Facebook allow access to the individual's account, granting them the ability to understand their social history and see their feeds. Or, if the government wanted to get inside an encrypted iPhone, they could have Apple (which would first put up a fight) unlock the device. Your average citizen can't access these back doors, but we don't seem to mind that they're there and that the government can.

THE US AND EU

After years of haggling and updating existing privacy laws, the EU created the GDPR—now considered the global gold standard. Other countries adopted this legislation, adding their own touches. Now, at the time of writing in 2022, twelve countries (including Canada) have privacy laws that are in line with the EU philosophy and are considered "adequate" by it. The US is not among them. Whether or not we eventually come into compliance with EU regulations, any American company that does business overseas will need to contend with both US and EU laws. In fact, the issue goes beyond the US and the EU.

For a while, US companies could go through a self-certification program called Safe Harbor. Once certified, they could transfer data over borders without violating any global regulation. But the program was brought down, indirectly, by Edward Snowden. When Snowden blew his whistle, he woke Europe up to the privacy issues at hand. Long story short, the EU no longer trusted the Safe Harbor program and shut it down, leaving thousands of companies wondering how to continue conducting their businesses and moving data overseas. The eventual answer for most companies was Standard Contractual Clauses (SCC)s.

SCCs

SCCs are exactly what they sound like. They're standard European contract clauses in twenty-eight languages which anyone can search Google to find and use. While a solution, these SCCs were notoriously complex and created many difficulties for international businesses, leading to the creation of Privacy Shield (Europe's second attempt at expediting its privacy laws for overseas companies) in 2016 with the goal of encouraging international business to flourish once again.

But, as any scholar knows, history repeats itself. Max Schrems, a privacy advocate and lawyer with a bent against US companies, started poking holes in Privacy Shield in July of 2020. After he exposed over 400 loopholes, the certification program was evaluated by the European Data Protection Board and found to be inadequate. In July of 2020, the European Court of Justice (ECJ) invalidated Privacy Shield. Companies were back to using SCCs yet again.

But soon, the SCCs were updated with more explicit (and convoluted) contracts. Companies transferring data to Europe had until December 2022 to complete evaluations of their cross-border data movement and comply with new laws. Any new contracts signed now must be the newer versions. There are other (simpler) ways to handle cross-border movement, but those are typically reserved for massive companies—cash is king, as they say. At the time of writing, a new round of rules is in the works, but if they're agreed upon, they're as likely as their predecessors to subsequently come under fire, and the old ones will be continually revised and updated keeping uncertainty alive and regulations ping-ponging across the Atlantic until the US adopts a view closer to the EU's.

THE FUTURE OF DATA LAW

We look at the past to understand the future: the world at large is going the way of the GDPR, with Canada, India, Brazil, and Japan already falling in line. America, on the other hand, is much more focused on capitalism. We Americans just like to build great companies and then make post hoc regulations when problems arise. While successful in the past, this mentality is the opposite of what we need in the digital age and could be our Achilles' heel in the world of data privacy and security. Just because it worked for seatbelts, doesn't mean it'll keep our figurative heads from going through the future's silicon windshield.

But simply "becoming EU compliant" isn't the best option for America. Our distinct philosophies have been ingrained in our cultures for so long that some American companies find their businesses simply cannot operate under the EU's privacy laws. The GDPR is the gold standard, not the absolute standard. America will continue moving closer to EU law while maintaining its own goals in commerce and governmental regulation. We have no choice. The future is coming faster than ever before, and the threats are real and multiple.

THE FUTURE OF DATA CRIME

In the same way that the Industrial Revolution introduced a new kind of impersonal and anonymous crime (and hence the murder mystery and Sherlock Holmes), the digital revolution has made possible a new (even more impersonal and anonymous) kind of crime that was unthinkable just a few years ago. Cybercrime is the unintended consequence of all our technological innovation; we couldn't have guessed at what dangers would arise from it. Formerly, if a mistake was made in data, you'd just get a new piece of paper and start over. Now with our automated efficiency, a mistake may mean your entire network comes down.

I (Justin) once participated in a tabletop exercise (a simulated breach) for a hospital in which the attack originated in their HVAC systems. They had smart technology that controlled all the heating and cooling in the hospital, a system accessible by the internet. You should have seen their faces when I showed them I could ruin their lab samples by cranking up the heat remotely. The laboratory is the lubricant that runs all hospital diagnoses and what allows them to treat patients. Should hackers have decided to access their HVAC systems (which wasn't hard to do), they could have held the entire hospital and its patients' lives hostage.

This hospital had never considered its maintenance staff to be part of its cyber incident response team. But after seeing how *anything* that connects to the internet is a vulnerability, you can believe they walked away with a different view of cybersecurity.

Digital information is, well, digital—you can't see it. Unlike the government's effort to spread awareness of using seatbelts, here the carnage isn't graphic and readily observable. But people are waking up to this reality. In fact, 2021 saw the first global summit on ransomware. The timing is not coincidental. I'll keep the history lesson brief.

During the nineties, just-in-time delivery was all the craze in the US. Then over the next thirty years, we outsourced our manufacturing to China. This worked perfectly in our service economy; inventory costs were low with this delivery system, and all our manufacturing needs were taken care of internationally. In a global pandemic, this delivery system becomes our Achilles heel and makes us vulnerable. How? There's no excess inventory to tide us over, and work comes to a halt while waiting on other countries to send us PPE and other vital business supplies, making us beholden to countries like China.

We've had such a focus on efficiency and lowering costs that we didn't consider the consequences and over time built an economic system that has now come to a screeching halt: nothing is free. Our advances in technology are repeating history in a different context. Cybersecurity is an offshoot of the pandemic and shipping crisis. Without backups and redundancies to our systems, one hack could take down a global economy.

Cryptocurrency is another direct result of innovation in the digital age. A logical progression in the evolution of money from representative to fiat to digital, the advent of cryptocurrency quickly created a large industry that decentralized finance. Now hackers don't need to steal money; they can ransom a company's data for a currency that is difficult to trace.

The Colonial Pipeline hack offers a final example of another vulnerability hackers exploit—the future's blurring of IT and OT. Colonial Pipeline had both an Information Technology (IT) and Operations Technology (OT) department. Their IT department did all the usual IT things—it stored data, managed their network, connected computers, and employees—while its OT department was where their control systems were. Gone are the old days where roughnecks would travel down the line with massive wrenches turning valves. As with all oil, water, and gas pipelines today, everything is operated remotely via Supervisory Control and Data Acquisition (SCADA). With the push of a button, the OT department can close or open pipelines a hundred miles away.

Obviously, digitizing this workload is incredibly efficient—less payroll and no more lost fingers—but it comes with a level of unforeseen risk. When IT and OT overlap, hackers break into more than data. They can control the actual functions in a business—stop or start machines, open and close doors, activate locks, and operate controls. A business owner whose physical infrastructure is connected to the internet is at risk of a hacker gaining the ability to operate or shut down any or all of it.

Business owners need to put cybersecurity on par with their marketing and start to develop a "Defense in Depth strategy," meaning there is no magic bullet or one piece of technology that magically solves their digital privacy and security threats. We'll cover this in much greater detail in Chapter 10. Here, it's enough to understand that—for all its marvels and efficiencies—the data technology, as with all new technology, is only as good as the intentions of those who use it.

CHAPTER SUMMARY

The chaotic history, patchwork present, and unpredictable future of data privacy and security regulation, coupled with the consistent (if no more organized) rise of cybercrime, is understandably intimidating to business leaders. As the most visionary tend to be optimistic and impatient by nature, teasing out the tedious regulatory details and investing in prevention lacks a certain sexiness. We get it. But it's also an enormous opportunity. Because the future—with all its wonders and the correlated complications we've just outlined—is coming for us all, whether we're ready or not.

So why not be ready? Anticipate the coming laws and regulations and enact customer-respecting privacy policies before you're forced to. Protect their data and yourself from foreseeable dangers before you're taken off guard. Be among the first in your industry to offer digital "seatbelts" and let your customers ask the competition why it's lagging behind and letting them down. Everyone will have to do what we're recommending eventually, but the positive psychological benefits will not accrue equally to those who wait until they're legally obligated to.

PART TWO

DATA COLLECTION AND STORAGE

CHAPTER 4
DATA COLLECTION

Even before they listed it as an area for improvement on her first performance review, Maryanne knew she lacked self-confidence. Having it pointed out didn't do much to improve it, but it wasn't the reason she lost her job. Maryanne was fired from Spectra Health for a trait she believed was one of her strengths.

Fresh out of college, Maryanne loved the health food and vitamin supplements company she worked for and believed in their products. She worked hard, learned quickly, and was thrilled (if a little unsure of herself) when she was asked to write the email announcing the company's new weight management program. She did multiple drafts and had a friend proofread for her. Finally, when her boss signed off on it, Maryanne went into the company's central CRM, considered who would benefit from the exciting news, and decided the answer was everyone!

Maryanne was fired for her enthusiasm.

A man who'd signed up for a webinar with the company's sports division was offended that the company thought he was fat. A woman who said she'd only supplied her email address to get a one-time discount phoned customer service demanding to know how they'd gotten her name. Worst of all, so many of the company's customers had unsubscribed from their newsletter, ticking the "I never signed up for this list" as their reason, that their email service provider flagged their account as a source of spam. Of course, it wasn't really Maryanne's fault. It was Sam's (and twenty-plus other sales reps, marketers, and execs).

Like many companies, Spectra Health didn't have a data collection policy so, whenever Sam, in marketing, wanted to conduct a survey to find out how people reacted to a company event, he'd go through the API integration and select every available field: name, email, age, income bracket, job title, and IP address. Had Sam ever been asked (and he never was) why he collected so much data, he probably would have said something like, "Why not? It's fast, easy, and cheap, and I might need it someday."

Most companies have a "Why not?" mentality about the collection and use of data, but it's badly outdated and needs to change. Companies that want to build trust with consumers will need both a mindset shift and new processes for how they manage the collection and storage of data.

THE MINDSET SHIFT

In today's business world, many marketing teams go and find whatever software that works best for them without notifying the IT department. They'll simply sync their customer data to a CRM, and no one else who uses that same tool will give a second thought as to how or why this log of information was created. The sales department then comes along and uses that stored data for their next campaign, and suddenly you have the same issue that sunk poor Maryanne.

Depending on the CRM being used, this problem can range from a minor issue to a disaster in the making. If companies collect and store home addresses, social security numbers, and other such personally identifying information, they need to understand how precarious their situation is, and not only from outside attacks. Companies need to regulate internally as well.

Your marketing team doesn't think about this. That's not their job. They just think it would be "so cool" to analyze their CRM and never stop to consider the data privacy implications. It's their job to just create invitations, send out sales emails, and get to know the customer base. Birthdays, names of children, and anniversaries of prospects are piled up in a CRM, and this data grows and grows. The reality is that no one ever decides to clear out the unnecessary data because "you never know when you'll need it."

I (Jodi) once consulted for a company that had five CRMs in place—and not because they needed to. They got one, didn't like it, got a second they liked better, then found a third they really liked. A separate part of the company got one of its own, and the fifth was introduced by the new CEO who brought his own pet system with him. The risks are obvious and increase with the multiple places data is held. CRMs are huge stockpiles of information, some of which have serious laws which pertain to them. Their loose nature and their users' instinctive desire to hold onto data create the potential threat of data breaches. But, as CRMs are invaluable tools in business, what data you're going to collect and how you'll collect it are too important to be left to chance or personal proclivity.

> Necessary mindset shift = Have a business reason for the data you collect.

THE PROCESS SHIFT

The shift from a "Why not?" to a "Why?" mindset needs first to be enacted in a company-wide policy that clearly spells out what data they will (and won't) collect, where, and how long it will be stored, and who will have access to it—all of which must be based on having a specific business purpose.

DATA COLLECTION

New Process: Data Collection

While recently looking for a babysitter, I was asked by a website for my child's full birth date (day, month, and year) before I could search for available sitters. Why did they need this? Why not just add an option to search for different grades or age ranges? What's the business purpose? Businesses ask for information that they don't need all the time. Think about the last time you went to a doctor's appointment and had to fill out paperwork. Likely, you were asked for your social security number. However, social security numbers are no longer needed by healthcare facilities; they just haven't updated their forms. Or look at what loyalty programs you're a part of. When signing up, did they ask you for your birthday to send you a celebratory coupon? If so, did they ask for the year or just the date? In case you missed it, they don't need your age to wish you a happy birthday.

The babysitting company was not practicing the most basic privacy principles of notice and choice*, which cover what you collect, how you use it, and who you share it with. A little extra thought put into what they were asking for would eliminate a customer's questions and build their trust. Because now I'm wondering how long this company was going to keep my data and what they were going to do with it. If customers understand why they are being asked certain questions, they are more apt to go along, trust your brand, and make a purchase. They'll feel more comfortable with how you will handle their data if your reasons for collecting are obvious to them.

> *Notice and Choice
>
> Notices tell people what you're doing and, once posted, need to be followed. Say what you'll do, and then do as you've said. Choice means giving people options about their data, such as the right to opt out, delete their data, or access their data (to name a few).

What to Do

Every company needs to think carefully about how and what data it collects and then articulate clear guidelines and procedures. There needs to be an easily accessible, internal privacy policy that explicitly states what can and cannot be done regarding data. Plans should be laid out in advance on how to handle never-before-seen situations. But, more than anything else, there needs to be thoughtful consideration of what data is being collected and for what purpose.

The first step in creating a new data collection policy is to set boundaries and guidelines and to create procedures for what data will and won't be collected based on the purpose of its collection. For example, in API integration, a company may need only a few data points: name, email, and customer ID. Collecting that specific data has a clear business purpose. But what happens often is that even though there's no real need for IP addresses, shopping cart values, browser types, etc., people will still grab any available data anyways because, again, "Why not?"

Most companies start from a boilerplate data collection policy, which is a great start, but these must be tailored because every company is unique. An e-commerce company needs different kinds of data and is exposed to different risks than a B2B organization. Even e-commerce companies differ in their marketing tactics and data collected on their sites. For example, some e-commerce companies will have a virtual fitting room and might ask for detailed body measurements. Another might just sell light bulbs.

Companies that have API feeds have different requirements as well. A company's privacy policy should reflect the type of data they're collecting. Understanding what data is being collected is also vital to being able to protect it from hackers (more on this soon), but every company's policy needs to comply with its customers' legal rights and safeguard customers' privacy.

Data Privacy

In creating the data collection piece of your new, business-reason-based policy, there are certain classes of data to which you need to pay particular attention: health, financial, and so-called "sensitive" data.

Health Data

While regulated by it, all health information is not completely covered by HIPAA. You can have a private conversation with an acupuncturist that isn't a licensed doctor and doesn't take insurance, or you can use a health app where you log your weight, blood pressure, and sleeping habits. Neither of which would be covered under HIPAA. This distinction is important because customers don't always understand the difference. While such business owners may feel the burden of HIPAA lifted from their shoulders, their customers may still view their health data as private and expect it to be protected. Even if your company isn't legally required to be HIPAA compliant, meeting customers' expectations is just good business and will help you earn their trust (or at least not violate it).

Financial Data

The HIPAA of the financial universe is the Gramm-Leach-Bliley Act (GLBA). Entities covered here include businesses such as credit card companies, car dealerships, and banks. That annual privacy notice that you throw away from your bank is a requirement they are fulfilling. By law, they must update you on what information they're collecting, how they use it, and who they share it with.

Businesses often look at the relevant laws in their industry and stop there. These laws only apply to specific transactional data, not *all* data. A company is still liable to meet certain requirements in data privacy before a person even becomes a customer—think marketing. This more general data is covered by state privacy law and is just as important.

Sensitive Data

When Americans think about "sensitive data," financial and health information come to mind. We are *very* protective of these types of information; hence all the regulations in those industries, but there are plenty of other kinds of information that can (and under certain laws is) considered sensitive and deserving of special protection. The (often-updated*) chart below illustrates both the diversity of what has been designated as sensitive and the discontinuity between laws.

> *You can find a regularly updated version of this chart at DataReimaginedBook.com for your easy reference.

PART TWO: DATA COLLECTION AND STORAGE

		GDPR	LGPD	PIPEDA*	PIPL	CCPA/CPRA	VCDPA	CPA	UCPA	CTDPA
	Racial / ethnic origin	♦	♦	♦		♦	♦	♦	♦	♦
	Political opinions	♦	♦	♦						
	Religious / philosophical beliefs	♦	♦	♦	♦	♦	♦	♦	♦	♦
	Trade union membership	♦	♦	♦		♦	♦			
	Health data	♦	♦	♦	♦	♦	♦	♦	♦	♦
	Genetic data	♦	♦	♦		♦	♦	♦	♦	♦
	Biometric data	♦	♦	♦	♦	♦	♦	♦	♦	♦
	Sex life / sexual orientation	♦	♦	♦		♦	♦	♦	♦	♦
	Past or present criminal convictions	**♦								
	Mail, email, and text message content					♦				
	Precise geolocation data					♦	♦	♦	♦	♦
	Personal data collected from a known child					***♦	♦	♦	♦	♦
	Social Security, driver's license, state ID, or passport number					♦				
	Account login, financial account, debit/credit card number			♦	♦	♦				
	Citizenship status						♦	♦	♦	♦

* "Sensitive information" is not defined under PIPEDA. However, the Office of the Privacy Commissioner has provided guidance that some personal information is regarded as being "sensitive."
https://www.priv.gc.ca/en/privacy-topics/privacy-laws-in-canada/the-personal-information-protection-and-electronic-documents-act-pipeda/pipeda-compliance-help/pipeda-interpretation-bulletins/interpretations_10_sensible/
** Processing of personal data relating to criminal convictions and offenses or related security measures based on Article 6(1) shall be carried out only under the control of official authority or when the processing is authorized by Union or Member State law providing for appropriate safeguards for the rights and freedoms of data subjects. Any comprehensive register of criminal convictions shall be kept only under the control of official authority.
***under 14
****in California: in combination with any required security or access code, password, or credentials allowing access to an account

Notice that the first three lines (racial/ethnic origin, political opinions, and religious/philosophical beliefs) are highly subjective, and not what most Americans would consider privileged. If your company collects any of the kinds of data listed above, you're going to need to create policies that comply with different laws in different jurisdictions.

Customers' Rights

New privacy laws include a laundry list of individual rights: the right to deletion (a request that data be removed), the right to access (learning what personal info a company has), the right to change (request to correct wrong information), the right to data portability (collected data is consolidated in a usable manner for a competitor), etc. Knowing how collected data fits into your company's business processes and systems will help you honor individual rights requests like deletion of data, access to information, and the ability to opt out of the sale of data.

Laws that govern these rights address an interesting question: who owns the data a company collects—the company that collected it or the people they collected it from and about? Again, there's no universally accepted answer, but most agree that consumers should have some choice about how data collected about them is used.

The chart below details some of the rights different laws have accorded to consumers. Most are self-explanatory. "Right to know" means companies are required to disclose what data they've collected on a person to that person if they request it. "Right to data portability" is perhaps the most interesting. It means, at least in theory, that you can take your Spotify playlists with you if you leave Spotify, your contacts if you drop out of LinkedIn, and your movie queue if you quit Netflix.

	GDPR	LGPD	PIPEDA	PIPL	CCPA	CPRA	VCDPA	CPA	UCPA	CTDPA
Right to know	♦	♦		♦	♦	♦	♦	♦	♦	♦
Right to access	♦	♦	♦	♦	♦	♦	♦	♦	♦	♦
Right to rectify	♦	♦	♦	♦		♦	♦	♦		♦
Right to delete	♦	♦			♦	♦	♦	♦	♦*	♦
Right to restrict processing	♦									
Right to data portability	♦	♦		♦	♦	♦	♦	♦	♦*	♦
Right to object to processing	♦									
Right to limit use of sensitive personal information	♦	♦		♦		♦	♦**	♦**	♦***	♦**
Right not to be discriminated against	♦	♦			♦	♦	♦	♦	♦	♦
Right to appeal							♦	♦		♦

* for data collected directly from consumer
** Opt-in consent required
*** Opt-out required
**** Controller may offer different prices, quality, etc. in limited situations

The right to delete is perhaps the most cumbersome. Typically, a company has to irritate a customer for them to exercise this right, but not always. If, for example, I gave a realty agency a lot of information while I was shopping for a home, once I'd moved, I might ask them to delete at least some of what they had on file about me.

Complying with these rights requires more than a purely technological solution. If a customer asks your company to delete their information, you need a process for determining who that customer is, that they are who they claim to be, and whether anyone else is acting on their behalf. Whose job is it to answer right-to-know requests? Who's responsible for that information that needs to be deleted is also deleted everywhere—in the CRM(s), Mailchimp, Excel, Salesforce, and in all the systems of every vendor who has it? Companies need a unified system of people, processes, and technology just to make sure they're able to handle these rights-based requests.

Breach Notification Laws

Every state, including Guam and Puerto Rico, has a data breach notification law requiring companies to follow certain protocols when a breach occurs. Each has its own flavor, but the standard operating procedure once protected data has been accessed or exfiltrated* is more or less the same between states. Upon discovery of a breach, the CEO of a company is obligated to notify the state's attorney general. Investigations are then conducted and potential class-action lawsuits follow. Between the trials and multiple state breach notifications, the cost can be exorbitant—an excellent example of the old saw *"An ounce of prevention is worth a pound of cure."*

> *Access—an unauthorized person has viewed data
> Exfiltration—an unauthorized person has downloaded data outside the company network

Imagine you had a business with customers across America and suffered a data breach*—you would need to comply with fifty different state privacy laws (not to mention Guam or Puerto Rico). Part of the damage analysis is knowing where your customers are located and what type of data has been exposed. Depending on the category, certain states may not require notification. Others will.

> *It shouldn't be that hard to imagine. Data breaches in 2021 topped the already-record-breaking year of 2020— by 17 percent![11]

---

[11] Chris Morris, "The Number of Data Breaches in 2021 Has Already Surpassed Last Year's Total," *Fortune*, October 6, 2021, https://fortune.com/2021/10/06/data-breach-2021-2020-total-hacks.

There's really no way around having to go state by state in this situation. While they share much in common, the severity of the consequences surrounding a data breach makes the variances in each state's law enough to review individually. There's no one draconian practice that solves this issue—which brings to question, why not have one federal law to comply with? While this solution is well known, it lies in a torrent of politics, complexity, and argument that will not lead to legislation any time soon.

The kind of data you collect (and thus expose in a data breach) impacts how your customers feel about your company. People tend to react more strongly to their social security number* being exposed versus their email address. Both are private, but companies need to consider the public's sensitivity to the type of data that's being collected, less they face litigation *and* further loss of trust. Remember, too, that even a "minor" leak of nonsensitive data can cause a tremendous amount of trouble for your customers. A stolen database of just email addresses can expose them to phishing campaigns or spam and you to their justified anger.

> *A social security number is always considered sensitive information under data breach laws, but not necessarily under general state privacy laws. As of this book's writing,

> there are only five comprehensive data laws in the US, and they don't all agree that social security numbers are sensitive data.

CEOs tend to think that their privacy policies should only concern what's relevant to a data breach, but according to the law, it's just not the case. Data breach laws are specific on what is defined as a "breach": data that was protected and is now exposed. But since California led the way in 2003, each state has brought its own nuances to what type of data is considered sensitive when there's a breach in security.

These comprehensive laws focus on transparency in the collection and use of data. For example, if a healthcare company, not covered by HIPAA, collects health data and wants to sell it, these laws may require that customers be notified prior to such an action. This makes sense as anyone would want to be given the choice to not do business with a certain company based upon what they would do with personal information.

Buying Information

Of course, collecting isn't the only way to acquire data. The marketplace for information is massive. Every time you make a purchase with a credit card, make a car payment, or conduct any other financial transaction online, your information is collected by someone (e.g., Bank of America, Capital One, etc.). Individually, this is not of much value due to it missing certain levels of information depending on what data was needed for that specific transaction (might have only needed gender and last name). But if the data is collected from a wide assortment of companies and locations, they can be combined to create a more valuable data set—enter data brokers.

Data brokers, like Equifax, build informational databases by purchasing data from businesses and using technology to comb through social media and create customized data profiles. With this process, data brokers create massive stockpiles of information on millions of customers, users, and businesses. The cross collation of data from multiple fronts fills in gaps of personal information, giving a fuller and more detailed picture of an individual or entity. While not every piece of information is comprehensive on an individual, enough data is aggregated across a vast array of inputs that the compiled data creates a digital profile of a person with disturbing accuracy. This wealth of knowledge is of great value to a marketer trying to target a customer base.

But if your company buys data (or even uses a third-party vendor who does), you need to be just as vigilant about what and from whom you buy as you do about what you collect. Two financial apps (Yodlee and Plaid)* that connect brokers, banks, and other financial institutions to mortgage applicants and their information have already been sued over the sale of data.

> *Plaid settled in 2022, but at the time of writing, Yodlee's case was ongoing.

Because they offered the convenience of collecting the plethora of documents and info needed for a mortgage application and then sending it where it needed to go, these apps were ostensibly great for the customers. But since many brokers' processes were built around apps that sold their data, users started to complain. They just wanted an efficient way to share documents with lenders and brokers. While Yodlee and Plaid honored their service agreement by sharing specific user information with specific companies, they were also sharing it with others which is how they got into expensive trouble for violating personal privacy and selling financial transaction data to third parties without adequate notice or consent.

If your company buys information, you need to be aware that some data brokers acquire their data in an unscrupulous manner. Some states, like California and Vermont, require people to register as data brokers to add transparency and help minimize this problem, but other states don't. Many CEOs get into trouble this way as they'll unknowingly purchase data that was acquired by less than honest means. In addition, if a CEO shares information with a third party who subsequently sells it, they can face consequences there as well. California already requires companies that sell or share information to give customers the option to opt out, and other states will likely follow its lead.

Business owners need to vet their sources and those with whom they share data and confirm that they are buying information only from reputable brokers and sharing it only with responsible organizations.

CHAPTER SUMMARY

My grandmother came of age during the Great Depression, and as a child, I used to love visiting her house. Every drawer, shelf, cabinet, and closet was a treasure trove. There were old buttons stashed in jars, books of matches in coat pockets, washed and folded tinfoil tucked between books, and mayonnaise packets wedged into every empty crack of the fridge. You could find absolutely anything anywhere in her house. Actually, you couldn't find anything. But you knew it was there. Somewhere.

PART TWO: DATA COLLECTION AND STORAGE

Too many companies bring Grandma's mindset to the collection of data. They scoop up everything they can the way she collected sachets of café sugar and bars of hotel soap and organize it, to borrow one of her words, willy-nilly. Companies that want to gain a competitive advantage by building strong, trust-based relationships with their customers need to shift their mindsets from Grandma's to one based on having a business purpose for the data they collect and retain.

They need to articulate and document this mindset in policies and procedures that account for the privacy of all data, paying particular attention to health, financial, and other sensitive data, and which comply with the wide range of laws that govern customers' rights, breach notification, and the purchasing of data.

Once you know what data you collect and buy, the next critical question is, as it was whenever I wanted tin foil at Grandma's: Where is it?

CHAPTER 5
DATA STORAGE

We'll start this chapter with a pop quiz:

- Are your marketing and sales departments using the same CRM?
- Are they using the same email system?
- How many systems do you have overall?
- Can employees source their own digital tools or is it all centralized?
- Is your data stored on multiple clouds?
- When was your last data inventory?
- Do you know what kind of data you collect?

Most CEOs won't even know *how* to answer those questions, which is the beginning of the problem (and why our consulting businesses have been successful!). Let's start with the basics of what data storage is.

Data storage is literally wherever data is located. This could be a database on the company network, an email on an individual laptop, a PDF in Dropbox, or a piece of paper in a file drawer. In short, "storage" covers wherever data is kept. We're going to focus on the two most usual ones: on a physical device or in the cloud. We'll look at both in turn, but first, it's important to understand how software tools and data storage intersect and why data storage is also a security issue.

SOFTWARE AND STORAGE

Anytime employees put customer data in an email, on their phone, or anywhere other than where a comprehensive data storage policy dictates it be stored, they create an additional instance of that data and another storage location. Unfortunately, most people realize this and happily go on sending emails and saving contacts on their phone, creating duplicate storage locations and additional points of weakness.

This is further complicated by the fact that often when companies change the tools they're using, whether it be new software, CRM, or whatever, the question of "What do we do with the old data?" never arises. Companies that don't have processes in place will either forget about the old data or keep it around—you guessed it—"just in case." We've even seen companies do this unwittingly because they didn't understand that their data didn't disappear when they stopped paying for it. They could no longer access it, but it was still there, and they were still responsible for it.

STORAGE AND SECURITY

Chapters 8, 9, and 10 will explore cybersecurity in greater depth, but relative to data storage, security is concerned with protecting three things: the confidentiality of data, the integrity of data, and the availability of data (CIA for short). And before you say anything, I didn't make that acronym up.

> Jodi: I know.
>
> Justin: Cybersecurity people aren't wannabe spooks.
>
> Jodi: I know.
>
> Justin: We're not all Spy Vs. Spy super high-tech road warriors of the information superhighway.
>
> Jodi: I know.

Justin: Just sayin'.

Jodi: It's still little James Bond, though, isn't it? *Cyberwarfare?* Just the word *cyber*.

Justin: Yeah, like an Aston Martin with modifications!

Jodi: Huh?

Justin: I'll make us martinis—shaken, not stirred.

Jodi: We have work to do.

Justin: Oh. Right. Understanding the principles of cybersecurity helps you see where the risks lie in data storage and how to better prevent breaches.

Confidentiality

Confidentiality is the assurance that data is only being accessed and used by those with the authorization or legal right to do so—which defines a data breach as: an unauthorized person gaining access to confidential information, thus breaching its confidentiality. When a hacker leaked Donald Trump's tax return on the internet, he broke into the accounting firm's servers and gained access to Trump's confidential data. Trump had the expectation that his personal information would be held in confidence by the firm.

Integrity

Integrity is about the completeness and accuracy of data. Avionics makes daily decisions based on a database of information. If their database were hacked and manipulated, the results could be disastrous; this is the next level of hacking. Imagine your company making critical business decisions based on invalid and manipulated data.

Availability

You can't use data that is unavailable, so availability is a much larger topic we'll touch on later. Here, we're concerned only with the denial of access. On a good day, access is denied to bad actors. On a bad day, the power of denial gets flipped—a hacker not only gets in and locks legitimate users out. Victimized companies, unable to operate their businesses without access to their data, pay large sums of money as ransom to regain it. Today's businesses need their data like the body needs oxygen, and a ransomware attack is as immediate and compelling as having your head held underwater by a Bond villain.

STORAGE OPTION: PHYSICAL DEVICES

The major advantage of storing data on networked company hard drives is that software exists that allows it to "live" in only one location, but to be accessed on and from any machine. The company owns the data, sees the data, and can wipe the data at any time. The downside, however, is that the same pathways that allow anyone access to anything make anyone able to infect everything with a single successful phishing scam.*

> *Phishing is a form of cyber attack in which hackers send an email or text lure which gets people to "bite" on a link that then "hooks" them by downloading malicious software that can do anything from destroying a hard drive to capturing keystrokes to leaving an open door to the network through which hackers can come in at their leisure and take (or break) whatever they want. (Both phishing and malware are covered in more detail later in this chapter.)

Rather than networked company hard drives, some companies opt to have employees house data on individual laptops. The usual preference, for companies that can afford it, is to issue company laptops and minimize the risk with policy and procedures. While this prevents one infected machine from contaminating the network, it's incredibly difficult to clean up unused data on so many unconnected machines. The number of data storage locations is equal to the number of laptops employees have, and every duplicate introduces complexity.

In the early days of the pandemic, however, many companies had to pivot so quickly that they didn't have time to buy laptops for all their employees and let them use their own. Now, they're seeing the consequences. Vital company documents may live only on a single computer that is poorly maintained or erratically backed up. When there's an issue, it can be very awkward to ask people to hand over their personal property.

The same situation occurs (and is only increasing) with the rise of the gig economy and the trend to hire even long-term employees on a freelance or contractor basis. Typically, a CEO will want to let these workers use their own machines just to have one less thing to worry about. But the increased convenience comes with a cost, and that is usually more privacy and security risk.

So pick your poison: do you want the security with some hassle or the convenience and the risk? So, before deciding on your approach, ask yourself the question, "Can I afford an incident?" Not to instill fear, but after seeing hundreds of incidents ourselves, the conclusive answer is no. Security breaches have increased 67 percent since 2014,[12] and 2021 topped the already record-breaking year of 2020—by over

---

[12] Kelly Bissell, Ryan M. LaSalle, and Paolo Dal Cin, *The Cost of Cybercrime*, Accenture and Ponemon Institute's Ninth Annual Cost of Cybercrime Study, March 2019, https://www.accenture.com/_acnmedia/PDF-96/Accenture-2019-Cost-of-Cybercrime-Study-Final.pdf.

17 percent.[13] With the trend continuing to worsen and a cost that includes the trust of your customers, prevention far outweighs the capital investment of supplying company laptops and phones.

STORAGE OPTION: CLOUD PROVIDERS

If you have your own network, you have control over data and increase your security. However, it comes with a significant cost to build and maintain it; and for many businesses, it's not their core function.

This makes outsourcing to third-party cloud-based systems highly desirable. With the headache of hiring IT workers, managing a complex system, and building up all this infrastructure, hiring a service is a no-brainer for most business owners. But simply off-loading this part of your business doesn't exempt you from the responsibility of protecting your customers' data. A cloud service has the same weakness any network does—it's only as strong as its weakest link. Without two-factor authentication (more on this later), the same hacker who phished an employee to access the company's internally maintained network would have the same luck getting into its cloud-based one once he had the employee's username and password.

---

[13] Morris, "The Number of Data Breaches in 2021."

Where's Your Data?

The best way to determine what data your company has stored where is to conduct interviews. Break your business down into its core functions (HR, finance, marketing, sales, customer service, etc.) and then delineate the data processing activities that take place within each. For example, the processing activities of the marketing department could be email marketing, digital analytics, and hosting webinars and trade shows. Within these processes, you'll want to inventory what data is being collected and where it goes.

During the interview process, speak to the employee(s) that know the most about the data flow for each process. Take note of where exactly they file what. During your interview with the marketing department, for example, you may find that your employees take the names and emails of website visitors, store them in Salesforce, and then start sending the prospects emails from their company Gmail account. Go through each core function and its processing activities systematically to get the full picture of your company's data flow.

For some companies, a good spreadsheet* will be a great start. For others, organizational software will come in handy. You can use it to send out assessments to your employees to review their job processes and data flow. You can also use software to conduct and steer the conversation with your department heads. As you move upstream, you can use targeted discovery tools that connect to your systems and help you grab what data exists. They're tremendously helpful for finding structured and unstructured data, but they only tell you what's where. Automated scanning tools don't help you understand why you have the information in the first place, what you're doing with it, or with whom it's being shared. This is why it's important to have the trinity of people, processes, and technologies at your disposal.

> *For a simple and quick way to get going, check out the free downloadable spreadsheet template on the *Data Reimagined* website DataReimaginedBook.com

The process of running a data inventory is like a doctor asking you to fill out a form detailing your medical history. It collects all known information and gives a 30,000-foot view of how "healthy" your business is regarding data.

Do this annually. Documenting all the data in a company and its flow—known as a Record of Processing Activity (ROPA)—is required under the EU's privacy laws. And as we've been discussing with California's leadership and the many other states' new legislation, increased privacy regulation has become the norm, making a data inventory even more crucial.

Although we know this may sound overwhelming, there are excellent data discovery tools available that can help companies locate data that might not be found in the interview process. These tools can find structured and unstructured data, which is great from a security perspective, even if it's not helpful for answering the privacy questions of why you have the information, what its business purpose is, how you're using it and who you're sharing it with. For that, you need the holy trinity—people, processes, and technologies.

WHAT TO DO

An annual data inventory is the foundation of a solid privacy program. Without it, you are merely guessing at what is needed in your company from a privacy perspective. But an inventory's benefits don't stop at compliance. They are three-fold. A data inventory can be instrumental in privacy law compliance (e.g., documenting the business purpose for using the data, writing privacy notices, and managing individual rights requests). It helps surface what's most critical to protect and supports recovery from cyber-attacks. And it provides a competitive advantage. While you probably won't want to brag that your company conducts data inventories at regular intervals, you can advertise how secure your company is and that you comply with US and global privacy laws like GDPR, if applicable. It shows your customers you care about their data, and this builds trust.

How Long Data is Stored

How long data should be stored is relative to your business. If Amazon deleted your shopping history after six months, you'd probably be upset. Amazon knows that people want to remember what they purchased two years ago—it's expected. However, does a pizza joint need your order from last year? Probably not.

Every company needs to think seriously about what data they need to keep and for how long. Some, in fact, do need to keep customer data on file indefinitely. But do they need to keep *all* their data? Remember the philosophy for data in business: for every data point you have, you expose your company to a data breach and elevate risk. Your doctor's office probably does need to keep most of your health data, but Zoom might only need some data for the length of one of its sessions. They don't need IP addresses on file consistently.

The rule here is the same: have a business reason for how long you retain the data you collect. Companies are continually over-collecting and over-storing data, creating unnecessary risk.

Companies do this because it's easy, cheap, and the data might be of use one day. But privacy laws are now being written that include specific language about data minimization, requiring you to collect as little as possible and keep only what you have a business reason for retaining.

When you don't need it, delete it.

These laws are meant to protect the consumer but can work to your benefit. If your company only needs 1,000 records, but you have been collecting everything you could and have 10,000 records, in the event of a data breach, your liability jumps by a factor of ten. Trust us as professionals that make this our business; you do *not* want that to happen. In general, keep as little data as possible in accordance with legal obligations and only where there's a business purpose. We ask our clients all the time, "Do you *need* access to this data?" If they don't, we get rid of it. Holding on to data gives you some level of responsibility for protecting it, because, if exposed, you could potentially have legal responsibility for it—which is counterintuitive to the pervasive philosophy of "collecting all you can."

CHAPTER SUMMARY

Companies need to know where their data is stored, and how its confidentiality, integrity, and availability are secured and regulated. That allows them to make informed and hard decisions about the relative advantages of storing it on the cloud, or on their company-owned networks and laptops. They need to keep track of what data is stored where by conducting regular (and regularly updated) data inventories and including how long they retain what as a part of their data collection policy.

Technology can help. There are terrific tools for finding and identifying stored data, for tracking privacy law compliance, and for data retention. Technology, however, is only as good as the people who use it, and of course, building trust with data goes well beyond what you collect and how you store it. Data is collected to be used, and how you use your customers' data is perhaps the most critical aspect in the building of trust.

PART THREE

DATA ACCESS AND USE

CHAPTER 6
DATA ACCESS

On the day in early December 2021, when Rupert Fuller's email was hacked, nothing happened. For ninety days, in fact, business at Full Tilt, his father's business where Rupert served as controller, went on as usual. As the person who handled the organization's many overseas wire transfers, Rupert lived an orderly life and went on about his business entirely unaware that hackers were quietly watching his routines and his schedule of invoicing and payments. On the fifteenth of December, he sent an invoice from Rupert@Fulltilt.com to the company's main French distributor, requesting payment by wire transfer for the previous month's purchases. On January fifteenth of the new year, he did the same. And again, on February 15, 2022.

In early March, the hackers went to work. They accessed Rupert's email program to redirect any correspondence to or from the French distributor to a dummy address and to hide any "failure of delivery" emails it generated. Then, on the fifteenth, they sent an email from Rubert@Fullltilt.com explaining that Full Tilt had changed banks and requesting that the French supplier send their payment to the new account. Rupert's counterpart replied to the email (never noticing the extra "L" in the hacker's spoofed email address), confirming the request, and then dutifully wired half a million dollars to Fuller's "new bank" in the United Kingdom.

It wasn't until Rupert's careful accounting system notified him that the French were uncharacteristically late with their monthly payment that the scam was discovered. But by then, of course, the money was gone.

There are two morals to this story. This first goes by an appropriately cynical name—The Zero Trust Model. Under this model (and sadly, it's a good one), you never trust any emailed request for funds. Assume all are false until verified through a secondary channel (e.g., a phone call*). Hackers prey on people's trust, and they're quite canny about our foibles. We rarely spellcheck a return email address and routinely respond to non-routine requests in as routine a manner as possible. The second moral is even simpler: use multi-factor authentication. Had the hackers needed an additional piece of information to authenticate their identity before reconfiguring Rupert's email, their plan would have been stymied before they sent the invoice.

*With deep fake technology, it's increasingly possible to spoof voices, so as the threat evolves, your channels of independent verification will need to as well.

Using data responsibly and in a way that demonstrates your respect for customers requires controlling who—inside your organization and outside of it—you allow access to what and the steps you implement to prevent unauthorized access.

IDENTITY ACCESS MANAGEMENT

Most companies already have some basic safeguards already in place to keep random strangers and "the bad guys" from gaining access to their network. You probably can't sit in your doctor's office, log in to their Wi-Fi, and snoop around their email, patient records, HR, or accounting software. They may give you a password to use their Wi-Fi, but it will not allow you access to files doctors and staff use for patient care. This is called Identity Access Management (IAM).

Identity Access Management involves identifying who is allowed to be on a network, authenticating their identity, and determining what level of access they have. The first question that needs to be addressed is how people establish their identities digitally. In other words, how will the computer know a person is who they say they are? How do they establish their identity?

PART THREE: DATA ACCESS AND USE

The most minimal form of identity is for employees to be given unique usernames and passwords. The next step up is multi-factor authentication, which requires a user to input multiple pieces of information to authenticate them to a system. This can include security questions (e.g., "What is your mother's maiden name?") or a code that is sent to your phone via text message. Some more sophisticated forms of multi-factor authentication are token-based (like Google Authenticator), fobs*, and biometrics (like fingerprint readers or voice identification).

At a minimum, if any websites or services you use offer multi-factor authentication, sign up for it—it's better than nothing. Even better, use a token system (Google Authenticator, Authy, VIP Access, etc.) if available. The basic concept is that certain levels of inconvenience for you will deter potential hackers. We are currently advising a CEO to only allow access to his multi-million-dollar cryptocurrency digital wallet with one specific phone and unique credentials. An inconvenience to be sure, but can he afford the risk?

*Fobs are devices that must be physically proximate to the user for them to log in. Think of the unique car fob that allows only you to open your car or start it. A data link connects to the fob and confirms it matches the user's

> credentials. A hacker who'd decoded a person's username and password would need to find that person and steal their fob for their other credentials to work. Not only does this form of authentication make remote hacking very difficult, it also rarely goes undetected. As soon as the legitimate user realizes their fob is gone, they know there's a problem.

Almost every company already has some form of access control in place around who is allowed to see financial information. Consider your HR department; not every worker has access to payroll, contracts, and revenue. Likewise, the same person rarely both cashes all the checks and does all the accounting. (And those that do often start magically getting richer.)

Access to data should be treated the same way money is. It should be based on what people need to do their jobs. The person who runs a doctor's scheduling and accounts receivable, for example, doesn't need access to patient medical histories, but the nurse who staffs telehealth does.

The solution is to parcel out permissions. This is not only respectful of privacy, it provides an additional layer of security. If every employee has access to any part of the network, one compromised password gives a hacker carte blanche. Segregating your network and isolating particularly sensitive parts provides privacy and forces hackers to depend on multiple mistakes being made before getting to a certain level of information in a network.

When setting up Identity Access Management, develop a system of rules with your IT department that govern who gets access to what data. Consider each job function in your organization and what it will need access to; that is, what critical information does the worker need to perform their specific duties? Then limit each person's access to the information they need to do their jobs. This process is even more critical if your company uses freelance or contract workers.

Freelance and Contract Employees

Imagine you need a bookkeeper. You go out into the marketplace and find an independent contractor; she has experience, she's responsive, has great references, works remotely—it's perfect. You do the onboarding, give her access to your accounts, and she starts handling the finances. Because such access is integral to her job, she has access to some very sensitive information—bank accounts, client lists, etc.—but she is trustworthy, and besides, you've got a good

IT team to handle the technical stuff should anything go wrong. Here's the issue: because your bookkeeper works from home, her laptop is not a part of your network. A hacker can easily access her machine and infiltrate your network.

By not hiring a full-time employee, you certainly save costs, but is it worth the potential security risk? Many business owners hand over their data to contractors, saying, "Get to work," simply glad to have some burden lifted off their shoulders. But contractors (and their computers*) need to be treated, from a privacy and security perspective, like full-time employees—the risk is just the same, if not greater.

Don't neglect privacy and security just because a contractor is small and independent. Freelancers need to be trained and taught their obligations to protect your data. Remember how nightmarish a data breach can be; exercise the same defenses and restrictions you do with your employees as with your contractors. Ask yourself what the worst thing this freelancer could do and then set up preventative measures to protect yourself. Can you create a better contract? Can you train them? Could you have them at least use a company email or a shared drive? Should you conduct a background screening? Long story short, be proactive.

> *While we recognize the extra expense (and hassle) involved in issuing company-owned laptops to remote employees, we recommend it. It's cheaper and easier than the many bad situations it can prevent. For contract workers, make sure you have a solid contract and that they're using two-factor authentication. You can't just turn over your data and hope. Do your due diligence.

I (Jodi) actually encountered a similar situation not that long ago. I was about to hire an excellent and highly qualified controller who, when I asked, assured me that she did, indeed, have a cyber insurance policy. Fortunately, I asked to review it before our work together got underway.

> Justin: Oh you did, did you?
>
> Jodi: Okay, I outsourced a review of her policy to Justin.
>
> Justin: And that was fortunate, wasn't it?
>
> Jodi: It was.

The "protection" her policy offered was all but useless. It didn't cover some of the most frequently occurring issues. Had hackers pulled the fake invoice trick they did with FullTilt, her insurance policy would not have protected us. Of course, every data situation is different, but it's important to make sure that you (and anyone with whom you contract) have a cyber insurance policy that covers the specific bad scenarios you might run into.

Revoking Permissions

I (Jodi) once did some consulting work for a law firm, and I'd probably still have access to their Google Drive and Dropbox if I hadn't explicitly asked them to revoke my credentials.

Companies need to include a thorough de-credentialing as part of their standard off-boarding process. Go through each system they had access to—email, CRM, main product system, whatever—and deactivate it. We know it sounds obvious, but if someone doesn't work for you, they shouldn't have access to your systems or network.

This is particularly vital if an employee leaves under less than congenial circumstances and goes to work for a competitor, or was working for you under a freelance contract. It's not that the freelancers are a threat, in and of themselves, but they're unlikely to have sophisticated security in place, and their computers make contact with lots of other networks.

PART THREE: DATA ACCESS AND USE

CHAPTER SUMMARY

The same mindset shift from "why not" to "why" pertains to the accessing of data as much as it does to its collection and storage. Have a business reason for who inside your organization is granted access to what data and implement strategies to prevent unauthorized access. Start by instituting a Zero Trust stance that assumes all emailed requests for banking or other sensitive data are bogus. Craft an Identity Access Management policy that uses multifactor authentication, parcels out permissions based on business purpose, and revokes them when appropriate.

CHAPTER 7
DATA USE

A city brought me (Justin) in to consult on the deployment of a new technology. One day, I went to lunch with the marketing team just to get to know a few of the people before I started complicating their lives. The head of Marketing was very excited about a new project the city had recently undertaken. They'd partnered with an app developer who was working with them to set up a city-branded app. Users—any city resident—could create an account and log in and find listings of civic-minded activities they could participate in—park clean-up days or food pantry shifts, for example. The app would track their participation and reward them for helping out with points they could then redeem at local merchants for discounts, free meals, and the like.

The city was excited about the opportunity, and the last step was the contract. The app collected users' names, emails, phone numbers, and a host of other information, and retained ownership of it. They could share it or sell it, and they could use it to text ads directly to users. Think about how you would feel if you signed up to help plant some trees in the park and get a free coffee, and instead your phone blew up with ads for everything from campsites to condoms. It would have been a public relations disaster for the city, creating the absolute opposite of trust—resentment.

Companies build trust with customers by using their data respectfully and communicating with them transparently about that use. Respectful use has three facets: policies, notifications, and preferences.

DATA USE POLICIES

As with data collection policies, data use policies need to be carefully thought through, clearly posted, and complied with. They also need to account for customers' rights and other legal regulations.

A good data use policy will tell customers what you intend to do with their data. If you collect an email address, is it exclusively for shipping updates, or are you automatically adding it to your mailing list so now they'll be getting regular updates and promotions? Did you collect their phone number so you can let them know if there's an issue with their order, or will you be texting them ads? Will you share the information you collected with third-party vendors? Will you sell it?

On Selling Data

The short version: don't.

The long version: If you're going to sell the customer data you collect, you're legally required to tell people of your intent and give them the option to opt out. Also, California (plus other states) may require you to register as a data broker.

The bad news: even if you have no intention of selling your customers' data, it's dangerously easy to do so accidentally. Under the California privacy law, the formal definition of the "sale of data" is long and specific. A rabbit ear quote would be "sharing information with another company for monetary or other valuable consideration."

If you share data with someone who subsequently shares it, you can be held responsible for its sale, *and* you are now in violation of your own privacy policy. Even if you carefully control your data sharing, many companies enter into joint ventures with other businesses to co-market or co-brand products. Under California law, this could be considered a sale of information, depending on how the deal is set up.

Another place people run into trouble for the inadvertent "sale" of data may be a familiar scenario if you've gone to a conference where, when you signed up, you were asked, "Can we share your information with our sponsors?" Or perhaps you weren't asked but started getting emails from the event's funders anyways. That's the deal; that's how conferences work. A sponsor pays however much—five, ten thousand dollars—to cover the event's costs and gets the registration list in return. If you never expressly mentioned that you wanted all these sponsors' marketing emails, your data—under law—may be considered "sold."

But perhaps where most people get surprised about what constitutes a sale of data is in digital targeting—building digital profiles of potential customers. Look at Facebook as an example. If I go to a website shopping for a new kitchen light and that website has a Facebook Pixel on it, Facebook now knows that I am looking to buy a light fixture.

Then next time I get onto Facebook, I see an ad for that kitchen light I was looking at. Even though I didn't get on Facebook to look for light fixtures, they know I'm in the market for new kitchen lighting. Facebook's point of sale to the light company is that they will use my data to deliver a targeted ad while Facebook itself adds my information to its ever-growing database—as your data and preferences are the real value! The "other valuable consideration" here is the ad tech and data analytics Facebook gets from its millions of pixels tracking people. Facebook didn't pay the light company for my information, but it's a transaction nonetheless, and the lighting company (and every company that uses Facebook Pixel) is probably guilty of selling my data—at least under California law.

THIRD-PARTY DATA USE POLICIES

Having determined how they will use the data they collect (including whether or not to sell it), companies need to communicate that policy to all contract workers or third-party vendors with whom they share data. Returning to a previous example, you wouldn't want the freelance bookkeeper you hire to use your client list to start soliciting new customers for her own business. Not only is that your confidential proprietary information, but it also constitutes a betrayal of your customers' trust. They didn't give you their email addresses because they're shopping for bookkeepers.

When you outsource a function, like payroll or customer service, you share private information with a third party who then holds onto it because they need it to perform their duties. When you do this, you have an obligation to make sure the information is only used for the specific reason you have given it. This all needs to be reflected in the contract; special provisions are included for specific industries under certain privacy laws.

Several years ago, the city of San Diego signed a contract with a vendor to install smart streetlights, which gave that vendor the right to sell or use the data from the smart streetlights to whomever or however they wanted. The data ended up being used to track people attending Black Lives Matter rallies and led to a public outcry, which claimed government deception by them touting the streetlights as innovation, not surveillance.

We've both seen this again and again: vendors with hidden data broker agendas. Data drives business, and people are not being transparent in B2B agreements, taking advantage of unsuspecting businesses and using their data. The backlash, when it comes, is inevitably directed, not against the actual bad guy but at the ostensibly good guy—the business they once trusted and to which they willingly gave their data.

Data Use Notifications

You are required by law to notify your customers about certain aspects of your use of their data, particularly as it pertains to privacy. I'm sure you've signed a HIPAA notice before or seen the annual credit card privacy notice as it falls into the wastebasket. These are examples of companies abiding by the privacy laws specific to their industries. These notices have to look a certain way and be sent at a certain time. What happens often is that companies either ignore this or write a privacy notice once and never look at it again. As business and laws change, privacy notices need to be updated.

The privacy notice is your opportunity to communicate with the customer. It's where you are supposed to be transparent, where you disclose what you collect and what you do with the customer's data—you know, *build trust*. Of course, you can just boilerplate a long legal version of the notice, but you'll miss an opportunity to show how you care for your customers. A nice, consumer-friendly, visual page that explains the ins and outs of your privacy notice does just that. Laws are making these notices longer and more difficult to read. A little effort spent creating a digestible summary can go a long way with your customers.

Many businesses copy the privacy notice from someone else's site because they don't want to get their own done professionally. The issue here is twofold. One, no two businesses are the same, and all aspects of a company's processes affect the others. One e-commerce business may have a loyalty program, and the other doesn't—the privacy implications change accordingly. Two, if you don't know what is supposed to be in the privacy notice, you are missing the point entirely.

Data Use Preferences

If you have children (or a spouse), you don't need us to tell you the difference between telling and asking. The data usage version of "Would you like peas or carrots?" rather than "You have to eat a vegetable" is the preference panel. By giving users a choice, you dramatically increase their sense of agency and thus their trust.

One of the best uses of a preference center we've seen was on the website of an accounting firm. When you arrived at the site, you could simply browse for information if you wanted to, but you could also create a logged-in experience. Here, they invited you to create a profile. The firm offered a variety of services—tax advice and preparation, auditing, bookkeeping, and employee benefit plan administration. In the preferences center, you could select the services you were interested in—perhaps only those related to doing your taxes—and deselect all the others. You could also tell them how you preferred to be contacted—by phone or email—confident that if you ticked the "email" box, you'd never get a call from them pushing their employee benefit plan services.

Almost any company with even a modest range of products and services would do itself a favor to implement this kind of prosocial pandering. Let people tell you what they want so you can give (sell) it to them. Had Box and Keg had this kind of opt-in preference center, I would have happily invited them to market tween furniture and décor to me. This is good for customers—they get more of what they're interested in and none (or less) of what they're not, and it's good for the business. Not only will customers stay on your lists longer, they're likely to buy more because you're better able to target your marketing—something most companies attempt to do with cookies.

Cookies

Not all cookies are created equal—chocolate chip is superior to snickerdoodle (but snickerdoodle is more fun to say), and oatmeal raisin is just an abomination. But we were talking about data, weren't we? Here, cookies are strings of code that record certain information about a user and are differentiated by their purpose rather than by chewy and crispy.

Non-advertising or "functional" cookies that keep you logged in or store items in a shopping cart don't require users to opt in. However, under the GDPR and the ePrivacy Directive, a user must give consent for advertising, website analytics, and other third-party cookies. This is why some sites will pop up what's called a "cookie banner," asking you to opt in. At present, in the US, they're not necessarily required, but it is a best practice, and consumers are starting to expect all businesses to have one. Fun fact: for any company doing business in California, a "Do Not Sell My Personal Information" link is likely also required in the footer.

Don't feel discouraged by this—it can be used to your advantage. A lot of companies use cookie banners as an opportunity to be humorous or use their branding and voice. Get creative! Starbucks has a great banner talking about how delicious regular cookies are. We've seen others that also mention the various types of cookies or just use really simple-to-understand language. It's important to also use your brand colors and tone in these banners, and don't forget about mobile; don't have them take over the whole mobile experience. We don't want to turn off customers by an improperly configured cookie banner that was meant to instill trust and transparency.

CHAPTER SUMMARY

Treat your customers' data the same way you'd treat them in person—with respect. Don't sell it and allow them to dictate their preferences. Create a transparent data use policy, communicate it to customers, and ensure that you, your contract workers, and third-party vendors abide by it. When done properly, this can yield a significant competitive advantage.

Data may be a commodity, but the people it represents aren't, and treating it as you would treat them builds trust and encourages them to share even more information with you—information that, when used responsibly and creatively, can be very good for business.

PART FOUR

CYBERSECURITY

CHAPTER 8

OF HACKS AND HACKERS

It's March 2020 and Declan Maher, the IT director of a medical company, has been told that in light of the pandemic, the company will begin working remotely. Although it's been a fully in-office organization for decades, Declan is tasked with getting over one hundred people up and running remotely. The C-suite puts immense pressure on him because every day that employees aren't working, the company loses productivity, and the quarterly report is coming soon. So he gathers up his team and works around the clock: nights, weekends, and pulling all-nighters. It all goes fairly smoothly, and people are able to work remotely relatively quickly; job completed. The execs are happy. Declan takes a well-deserved night off and eats too much pizza.

But speed meant shortcuts were taken by Declan's team—you can't blame them with the higher-ups breathing down their necks. "Little things" like multi-factor authentication and Identity Access Management got skipped. These additional steps, while good for security, take time to implement, and Declan knew that time was of the essence; the company's existence—and consequently his job—was on the line. He put these "nice-to-haves" on the back burner and, as the pandemic lingered, the more urgent issues involved in taking (and keeping) a large company fully online habitually took precedence over his backlogged to-do list.

Nine months later, a threat actor came sniffing around the company's network and saw that no multi-factor authentication systems were in place. He easily brute-forced a password of one of the many remote workers and soon had complete access to the company's network at the domain level. (Because the IT department was swamped in "How do I get to my CRM?" emails, they never got around to setting up access levels). The hacker did a bit of reconnaissance, prepped his malware, and a few months later, when Declan went into the office on a Saturday to do some catch-up, he booted up his machine and was met with a message: "Good morning! We've encrypted your network, and your data is no longer available. Please contact us at this email, and we can talk about how you can get your data back. Have a great day!"

Now hackers are demanding $1.3 million in Bitcoin to ransom the company's data, and the reaction toward the entire IT department is hostile. "How did this happen?" C-suite demands, "After all, you're in charge of our digital security." But in truth, Declan was just responding to the leadership. "Get this company running remotely as fast as possible" were the words Declan remembers. The business executives didn't understand that this quick shift to remote work would undermine data security and that this ransomware attack was a time bomb whose fuse they'd lit themselves.

This is the point in the story where I (Justin) get the frantic phone call—desperation thick in their voices. And for good reason. If it's a small company that's just been hit, its odds of staying in business are around 40 percent.[14] Nonetheless, I dive in. I start by asking what data they keep and where they keep it. Typically, they will have some vague idea, but particularly post-pandemic, data is rarely where the guy I'm speaking with thinks it is. After some digging, we find data in many places other than its designated locations. I confirm the data breach and deliver the bad news—they need to give a breach notification (which is too often met with a confused "What's that?")

Most companies have policies that spell out where data is supposed to go and be stored—customer financial data goes to server number four, for example. But once I've finished the forensics on their

---

[14] Joe Galvin, "60 Percent of Small Businesses Fold Within 6 Months of a Cyber Attack. Here's How to Protect Yourself," *Inc.*, May 7, 2018, https://www.inc.com/joe-galvin/60-percent-of-small-businesses-fold-within-6-months-of-a-cyber-attack-heres-how-to-protect-yourself.html.

machines, nine times out of ten, I've found individual employees have downloaded customer information and the hacker has accessed their laptops. Companies never consider that sensitive information would be on those laptops because employees are not supposed to access that data—too little too late. Now they need to give notice and bring on the "rogue's gallery of plaintiffs," as I say.

It is extremely important to know what and where your data is. Data breach notification laws don't focus on how sensitive data is accessed, only that it's been exposed. That means even if an employee's laptop is stolen (and not password protected), any sensitive information on that laptop has been made vulnerable, and notifications may need to be sent. Few people understand until they've been through it, how awful a breach notification can be. Depending on the state and type of data exposed, you may need to inform the attorney general and become the subject of state and FTC investigations for deceptive trade practice, class-action lawsuits under the CCPA, and more.

The common denominator in these ransomware attacks I've worked on is that the C-suite folks had no idea how their networks operated. They didn't realize their employees can, and are, downloading unregulated software, company info, and playing esports all on the same machine. I'm always amazed at how little business owners know about their IT systems, which are basically the heart and bloodstream of their companies. One hack can ruin a hospital, halt a company's operations, or blow a pipeline.

How much of Declan's story sounds familiar? Cybercrime increased 569 percent over the course of the pandemic as companies did exactly what we've just described.[15] Are you feeling a cold sweat about some of the choices you made during the pandemic? If you are, that's good; you're starting to see the reality of cybersecurity. Let's dive in.

MEET THE HACKERS

I've found it helps CEOs understand cybersecurity better to become familiar with the psychology behind the threat. Accordingly, I group hackers into four broad categories: Activists, Nation States, Bandits, and Hot Shots.

Activists do their illicit work to make a social point. Think of an environmental group that would want to hack BP to make a statement or, say, the political impact of releasing a political campaign's internal emails.

---

[15] INTERPOL, "INTERPOL Report Shows Alarming Rate of Cyberattacks During COVID-19," press release, August 4, 2020, https://www.interpol.int/en/News-and-Events/News/2020/INTERPOL-report-shows-alarming-rate-of-cyberattacks-during-COVID-19.

PART FOUR: CYBERSECURITY

Nation-States have a different agenda with their hacks—addressing you, China—espionage, surveillance, stealing trade secrets, and the like. Instead of holding a system hostage, they will use malware, phishing, etc., to, perhaps, find the latest plans for the next-generation fighter jet or drone and steal it, and you may never know they were in your system.

Bandits hack for the money. They've stolen credit card numbers, hacked cryptocurrency, and held data or functionality hostage. Once hackers breach a system, they can sell pretty much anything they gain access to on the dark web.*

> *Think of the dark web as the seedy underbelly of the internet where, in some poorly lit alleyway, there's a guy in a trench coat offering credit cards and social security numbers. It's an entire underground economy where anyone can buy Facebook passwords, malware, ransomware, and more—a one-stop-shop for all things digitally nefarious.

Hot Shots do it for the status. In the body-building crowd, more muscle is what commands respect. In hacker culture, your "muscle" is your ability with a computer and what you've hacked. Creating a better hack is social currency. Bragging rights are what gives a hacker his name. In a lot of ways, hackers are similar to the CEOs they torment. A CEO may show off their success with a paycheck or car; it's a similar personality in a different context.

It's this last class of hackers that gives most CEOs their sleepless nights. Your company may be attacked for no reason. It can have nothing to do with money or business practices. It could be as simple as a hacker "flexing his muscles" and earning some social cred. Not every hacker has a criminal mindset. I once had one client who was hacked for just $5,000. The hacker could have demanded a much higher ransom, but he only wanted to pay for his college semester (and confessed to feeling bad about what he'd done).

But here's the thing about people who are good with computers—they're typically folks who were the "nerds" or "geeks" back in high school (or they're actually still in high school!). Now with such dependency on technology, they find themselves the only one of fifty employees who understand a business's network. They recognize the power of this but feel like no one else does. It's a way of exerting power and demonstrating skill. And they feel nothing but scorn for any system they can crack. To many hackers, busy CEOs (who make way more than they do) aren't victims; they're idiots. Some even see themselves as teaching a valuable lesson about taking security seriously.

MEET THE HACKS

The types of cybersecurity attacks are as varied as machines are complex. The creativity of those who made the first machines is matched by those that hack into them.

DOS Attacks

A denial of service attack (DOS) is when a hacker directs so much traffic to a website that it shuts down, unable to handle the volume. It's a bit like tripping a circuit breaker by operating too many appliances that are plugged into the same outlet.

Ransomware

Boiled down, in a ransomware attack, a hacker encrypts a network and demands payment in exchange for the decryption key. In our real-world analogy, it would look like someone putting a padlock on your door and selling you the combination. With the average ransom demand increasing 144 percent in 2021 to an average of $2.2 million and the average payment rising a full 78 percent to $541,010, is it any wonder that this is a particularly popular and profitable form of cybercrime? In fact, it's spawned its own secondary industry on the dark web—Ransomware as Service—complete with customer support teams to help you with decryption once you've paid up!

Phishing

Phishing—also known as spear phishing or whaling—usually comes hand-in-hand with wire fraud. A hacker will impersonate a vendor who sells a service or product to an unsuspecting business. Once the business wires the hacker money, the hacker disappears—as does your cash. You may have seen this personally if a "Nigerian prince" has ever asked for your help. Phishing may also involve emails or texts that look legitimate (and may even appear to be sent by someone in your company) and that include a link that, if clicked, downloads malware* onto your computer.

> *Malware is malicious computer code that can do a wild variety of things. As one example, there is malware that allows someone to watch all your keystrokes. If you were to login into a website or your email, a hacker would see the username and password you used, allowing them to access everything you can. If this happened to the administrative assistant of the CEO, a threat actor would have access to all the CEO's emails. Other forms of malware allow hackers to harvest your login credentials, giving them access to everything in your inbox.

In advance of a presentation I'd been invited to give, I emailed participants asking for their input on a chart I was considering for inclusion in my talk. My email included a link that took them to a website where they were asked to enter their email address to view the proposed chart. Of course, the website was bogus—designed to see who could be "phished" into giving up their credentials. Of the 120 who received the email, 57 percent clicked on the link. Of that group, 37 percent submitted credentials. And I had the data I wanted for my chart. Had I been a hacker, I only would have needed one person to fall for the same thing. Needless to say, I saw some sheepish looks from that audience, but it got their attention! How many of you would have fallen for the same trick?

Viruses

A computer virus is a specific kind of malware designed to quickly replicate and spread from device to device. This is very similar to getting a cold, sneezing, and having the virus particles from your sneeze land on the person next to you, giving them your cold. They, in turn, might sneeze and infect more people. In the computer world, think of clicking on an email attachment that contains a virus that gets into your corporate network and rapidly spreads to connected devices and servers.

Spoofing

Spoofing is what happened to Rupert back in Chapter 6. In this kind of attack, the hacker finds someone else's email and pretends to be that person. For example, if we were in business together, a hacker could send you an authentic-looking email that for all intents and purposes seems like it's from me. The email might say that we are changing banks and ask you to send invoice payments to a new bank account (the hacker's vacation fund).

SMS Scams

Losing your mobile phone these days is on par with losing your wallet or purse. If a hacker gets your SIM card*, they can change your usernames and passwords and receive your SMS security texts. They don't even need to steal your phone to access your SIM card. All they need to do is call your mobile phone provider and convince them to transfer your phone number to a different SIM card. This is easier to do than you'd think. Mobile carriers ask standard personal questions to authenticate the identity of callers, and if hackers can pick up this information from your social media postings, they can "prove" they're you. In 2021 alone, the FBI received over 1,600 SIM-swapping complaints with losses in the many millions.

Pro Tip: Call your mobile service provider and put a freeze on porting your phone number which can only be unfrozen by providing a password you *don't* post on social media.

> *A SIM card is what links your phone number to your phone (and you).

Man-In-The-Middle Attacks

Our cell phone has become a mobile command center which makes it a prime target for this kind of attack. Man-in-the-middle attacks have three players: the person sending the information, the person who is intended to receive it, and the hacker who intercepts it. Imagine you connect to public Wi-Fi at the airport. How can you know the network you've just logged into is legitimate and not one created by a hacker? If you connect, the hacker gains access to your phone, maybe even to the username and password you use to connect to your mobile banking app. They now can do anything you can, including transfer money and make purchases on any website where you have a credit card saved.

Man-in-the-middle attacks can also interpose themselves between you and a legitimate website by creating remarkably accurate fakes.

You think you're on your bank's site, but you're not. You're really communicating with the hacker.

Social Engineering

Social engineering isn't a hack, but a method of enticement that creates an opportunity. The digital version of the classic conman, it's any type of ruse that takes advantage of human weakness or social custom to lure you into taking an action that hackers then exploit.* Some of the most common are: "Congratulations! You've won a $100 gift card from Amazon or Wal-Mart or any place you're likely to have been in the past week" and "Reminder: your warranty is about to expire, renew now before prices go up." During the pandemic, lots of people received phishing emails from hackers posing as charities looking for donations to help people affected by the virus, preying on people's empathy to get them to click on an attachment. Pretty heinous stuff.

> *With 95 percent of breaches attributable to human error,[16] almost all rely to some extent on social engineering.

---

[16] Devon Milkovich, "15 Alarming Cyber Security Facts and Stats," Cybint Solutions, December 23, 2020, https://www.cybintsolutions.com/cyber-security-facts-stats.

CHAPTER SUMMARY

Activists and hot shots, nation-states, and bandits may have different motivations for what they do, but the damage they can do is immense. Between DOS attacks, ransomware, viruses, spoofing, SMS scams, and man-in-the-middle attacks, hackers inflict $2.9 million of damage to the global economy *every minute*.[17] And that price tag doesn't factor in its non-monetary costs. (We'll discuss what else is at stake in the next chapter.)

Hackers are brilliant, creative, and highly motivated. They're also utterly ruthless. That's the threat we're all facing. But what makes us vulnerable?

---

[17] Joel Johnson, "What Costs $2.9 Million Every Minute and How to Protect Yourself From It?" *Forbes*, August 11, 2020, https://www.forbes.com/sites/joeljohnson/2020/08/11/what-costs-29-million-every-minute-and-how-to-protect-yourself-from-it/?sh=4745c9687764.

CHAPTER 9

WHEN VULNERABLE IS A BAD THING

There's a charming series of questions that claims (with some significant clinical support) to make people fall in love.[18] There are thirty-six of them grouped into three sets which are designed to promote closeness by encouraging participants to share progressively more intimate details of their personal stories, hopes, and pain. Brene Brown has made a career out of encouraging people to be vulnerable, making research-based claims about its trust-creating powers.

But this is data, not dating.

---

[18] Daniel Jones, "The 36 Questions That Lead to Love," *The New York Times*, January 9, 2015, https://www.nytimes.com/2015/01/09/style/no-37-big-wedding-or-small.html.

Here, as we've been at pains to establish, safeguarding rather than freely sharing information is the key to building trust. Here, "vulnerability" really does mean "weakness."

THREAT VECTORS

The first step in protecting yourself, your business, and your customers from the hackers and hacks we detailed in the previous chapter is understanding how the bad guys get in. There are four different threat vectors: public-facing systems (e.g., email and websites, covered in Chapter 8), Wi-Fi, company insiders, and the supply chain.

Wi-Fi

If you're out working in a coffee shop using their Wi-Fi, a hacker at that table in the corner can see everything you're doing. If one of your employees boots up their laptop at the airport while waiting on a flight and connects to the airport's public Wi-Fi, they've made company data vulnerable. Public Wi-Fi can be hacked, and any laptop connected to that Wi-Fi is at risk. Even if the link is broken (the user shuts down their machine), with all the remote abilities nowadays, a hacker only needs to have stolen your login credentials to later access your system remotely from anywhere in the world.

Think of this like Legos: Wi-Fi is a Lego piece that can connect to a laptop (another Lego piece). A laptop may already be connected to other Lego pieces like a company's network or CRM. If a hacker can breach the first Lego piece (public Wi-Fi in this case), he can access whatever else is connected—it's a chain of Lego pieces. While interconnectivity is what makes networks (and Legos) so flexible, efficient, and appealing, it's also its biggest risk. That and bare feet.

Insider Threats

The third threat vector (which accounts for 60 percent of cyber incidents[19]) comes from inside your own company—people to whom you've given access. Most typically, it's a disgruntled IT worker or someone who was fired but whose credentials weren't revoked, and they can wreak all kinds of havoc.

Recently, a client I worked with had hired a new employee who reported it took a full week for her previous employer to cut her access to their network. We had a good laugh about all the things she could have done if she'd had a bone to pick with anyone there. Or if she had simply been greedy and unethical enough to profit from their carelessness.

---

[19] "Insider Threats Are Becoming More Frequent and More Costly: What Businesses Need to Know Now," ID Watchdog, accessed June 30, 2022, https://www.idwatchdog.com/insider-threats-and-data-breaches.

A Special Kind of Insider Threat: Governance

I'll take a quick detour to explain a special kind of insider threat called "governance." The easiest explanation is to look back at Declan's story from the start of the last chapter. Here, the insider threat came from leadership who quickly pivoted its business to have its employees start working remotely. Poor governance creates a threat not through hostile action, but from passivity. C-suite managers, out of ignorance or neglect, are often their own worst enemy when it comes to data privacy and security.

Supply Chain

The last, but certainly not least, threat vector comes in your supply chain, a source as dangerous as the other three vectors combined.* It was the entry point for the SolarWinds hack, which is more famous, but we'll use the less publicized story of Target's 2014 hack because it's also a great demonstration of my Legos Theory of Cybersecurity. Target wasn't hacked directly; its HVAC vendor was.

> *Over half of organizations have experienced a data breach caused by a third party.[20]

---

[20] "51% of Organizations Have Experienced a Data Breach Caused by a Third-Party," *Security*, May 7, 2021, https://www.securitymagazine.com/articles/95143-of-organizations-have-experienced-a-data-breach-caused-by-a-third-party.

The company was a heating and cooling systems specialist in Pennsylvania, with contracts to perform maintenance and repairs for many of the local retail stores. Its network was connected to Target's network. There was no Identity Access Management with respect to the HVAC vendor's access to the Target network. If you have read this far, you know that means the hacker could access and move around Target's entire network. And they did, specifically targeting the portion of the network used to manage Target's POS (point of sale) data from which it then stole tens of millions of dollars' worth of credit card information.

Target is a huge organization and hadn't skimped on software designed to detect such unauthorized access. Unfortunately, the IT workers who were monitoring the network took the warning signs as false positives attributable to the busy holiday shopping season and ignored them. Target had the right technology but lacked the skilled personnel to use it, making it just a fancy, unused, and *expensive* piece of software.

Another example of supply chain vulnerability is the SolarWinds hack. SolarWinds is a software company that provides cloud-based management tools for monitoring a company's network. One such tool, called Orion, had privileged access to another company's network to obtain system performance and log data. This and its widespread use by many Fortune 500 companies and the US government made it an attractive target for hackers. After all, why try to hack fifty individual companies when you can attack one common supply chain vendor and get access to all of them?

The SolarWinds hack began in September 2019 and was not identified until December 2020. A mere fourteen months! But in that time, more than 18,000 SolarWinds customers unknowingly installed the malware that allowed hackers to impersonate legitimate users on the network of every organization using Orion. This hack illustrates how using a cloud software provider creates a centralized point of failure for so many organizations and how our interconnected systems make the supply chain both so vulnerable and so enticing to hackers.

If any of the vendors in your ecosystem has a data breach, it will impact your company. Because every company you are connected to affects your business, you need to attend not only to your own but to their security as well. Even though you're not technically responsible if one of your vendors is ransomwared, if it's your fault in the eyes of your customers, it's your problem.

In another Lego chain disaster in which I (Justin) was involved, a professional services company called me in a panic. Pros 'R Us (not its real name) was pulling in $30 million a year—a sizable company, but not *too* big—that provided services to other companies who wanted to outsource the back-office functions of their business. Pros 'R Us hadn't been subject to a ransomware attack, but the hosting provider where they stored their customers' data had been.

When the hack happened, Pros 'R Us was held liable for the data breach because their clients had contracts with *them*, not the hosting company. Pros 'R Us had excellent security. They hadn't been hacked, and the hack wasn't their fault, but it was still their problem. I read over the contract they had with the hosting provider with a sinking heart. The hosting company's data breach liability to us was a total of $10,000. My client's liability to their customers? One thousand times that.

It was an uncomfortable conversation, but—and forgive me for going all *lawyer* for a moment—that's what happens when people don't read the contract. Fortunately, the incident didn't put them out of business. Since then, however, they've had several more breaches as a result of customers not paying attention to cyber. The supply chain is a *chain*. Legos don't care whether they're stacked above or below another brink. They're still connected.

Likewise, Pros 'R Us is connected to its vendors and its customers, many of whom are small healthcare practices. Doctors are famous in the IT world for their unwillingness to take instruction and, consequently, for being highly hackable. Pros 'R Us is connected to

Legos that keep getting infected, and it's causing serious problems. In fact, I advised them to rethink who their customers are, or at least to start requiring their clients to have better security because if Pros 'R Us continues to make insurance claims, soon they'll be uninsurable; and how will they do business then? They wouldn't be the first company to be put out of business by hackers.

A Special Kind of Supply Chain Threat: The Cloud

Many businesses like to put their information on the cloud. Sure, it's convenient, but most cloud services aren't just one vendor. That cloud vendor may have other vendors providing services like remote security or remote login. This is perfect for threat actors. A cloud has many clients gathered in one spot with several points of access through its multiple vendors—a gold mine for hackers. The Kaseya hack is an example of what can happen when hackers cash in on it.

Many companies outsource their IT to a cloud-based managed services provider (MSP). These MSPs integrate offerings like Kaseya's software that allows the MSP to remotely manage the IT network of its customers on the cloud. Similar to SolarWinds, once the hacker was able to breach Kaseya, they gained access to over 1,000 MSP customers where Kaseya was being used. This time, the hackers deployed ransomware to these MSP customers. The supply chain mixed with the cloud proved lethal to the companies who were disrupted when a vendor to their outsourced MSP was hacked. Do you think about all the third parties that are now integrated into the technology products and services you buy? If you purchase cloud services, do you ask who the third-party vendors are and inquire into their security hygiene? You should.

It seems the cloud builds an unvirtuous cycle when it comes to security. Companies that want to protect their data but don't want to build the infrastructure congregate in one place. This makes them an even bigger target and endangers their data all the more. Centralizing anything, no matter what it is, creates a common point of failure and an attractive target. And there it is again, the enduring theme of this book: efficiency and cost-effectiveness create security risks. Freedom or security—a debate as old as whether to burden yourself with a spear on the way to the watering hole.

THE STAKES

As we've already seen, activists, nation-states, bandits, and hot shots launch DOS, ransomware, phishing, and spoofing attacks on businesses—damaging their reputations, destroying consumer trust, and inflicting significant financial and operational damage. But that's not the extent of the harm they can do. I knew of a doctor who, because of a ransomware attack, had to shut down his practice and, sadly, ended up committing suicide. The private lives of several famous people have been made public without their consent, and the harm to their reputations and careers is hard to calculate—not to mention the emotional suffering it must have caused. Nor do I think it's overstating things to say we've already seen hackers change the political course of our country.

Hacks of elections question the integrity of the election process and undermine the very foundations of democratic society. Remember when Russia was creating all those fake Facebook accounts? That's real political harm. In 2014 and 2016, Cambridge Analytica was able to collect data and learn enough about individual users to create carefully targeted ads that changed people's political views. This isn't just a problem for the US. Other countries (among them Kenya in 2013 and 2017, Malta in 2013, and the UK in 2016) have also seen their democratic systems manipulated by hackers seeking to undermine them.

This is where a lot of my passion comes in data security: the harm being done to our country. If someone can subvert the right to vote and people believe they are no longer being heard, it damages society. There's no way of holding people accountable for creating or propagating misinformation on the web, and it's making people distrustful of even legitimate and responsible sources. With no legitimate, trusted, external referent, people fall back on their own biases to the detriment of us all.

Worse, in the future, I anticipate a range of other kinds of damage hackers will inflict on people, businesses, and societies. The data that we worship, the highly sought-after asset that we base our decisions on, is what makes us a target to those that would do us harm. In helping ourselves, we open ourselves up to more risk. Our ever-increasing development of technology, which begets more technology, demands more convenience and efficiency, enabling threat actors to create more inventive and damaging ways to use it against us. Imagine getting on your first autonomous drone passenger flight. (The day will be here before you know it!) If the data for its instruments has been hacked and manipulated, your life and the lives of the other passengers are at risk. Or think of a hacker that rewrites all the known allergies of a hospital's patients, demanding ransom for the correct data. What will the hospital do? Risk giving a patient medication that could kill them or pay up?

Little thought is put into how technology will affect culture, law, economics, privacy and security, or health and safety in the future. We are a people of innovation and action, and that characteristic is outpacing our policies—most notably in the realm of privacy. Here's a great example: say a neighborhood wants to deter break-ins and to be able to apprehend and prosecute offenders more quickly. So they unanimously decide to install cameras on every home, making any would-be criminals easily identified. Sounds great, right? But what happens when the "neighborhood watch" is connected to other cameras around the city? Between facial recognition technology and the vast network of cameras, digital profiles of individuals can be created to track their every movement in real-time all over the city—any Orwell fans out there?

It's already happening. Without even realizing it, we are building a surveillance state—think about how ubiquitous drones and autonomous vehicles will be in the next ten years. This array of technologies can be used to build better customer profiles for marketing or, more sinisterly, for harassment; consider what a jilted boyfriend or girlfriend could be capable of if they could easily geolocate their ex at any time.

There have always been unintended consequences when new technologies are deployed. Progress always has a cost. Autonomous vehicles, drones, and cryptocurrency will be no different. While the threat keeps evolving, people are fixated on the perceived benefits and wave off any warnings with, "We'll just figure it out later." This is the exact opposite of the approach that's needed. The issue is like a virus, evolving and becoming more complex, day by day. The time to act is now. Saying all this, I feel a bit like Noah telling his neighbors that the world is going to flood, but the consequences are real *and abundant*. Too often people realize the stakes only after they suffer the consequences of their data privacy and security choices.

THE IMPEDIMENTS

Despite the high stakes and the steadily mounting toll of damage done by hackers, we continue to exponentially increase our connectedness and our risk. This trend will continue and become worse until companies wake up and decide to make cybersecurity a strategic business enterprise risk, on par with their marketing and perfecting their services and products. Why hasn't it happened yet? For the same reason you'd rather buy a new sofa than replace your insulation—maintenance isn't exciting and growth is.

Maintenance

No business owner after getting a new $10 million contract wants to hear, "Now's the time to add that $5 million security software." Maintenance isn't sexy, especially data security maintenance. This isn't the new paint job on your Porsche; it's the oil change.

Think about all the problems with our physical infrastructure in the US—airports, roads, bridges, highways, you name it. Our digital infrastructure is also in disrepair, and for the same neglectful reasons. Even worse, digital infrastructure is unseen—*out of sight, out of mind*. One benefit of hackers is that they have provided a bucket of cold water to wake up those in charge of digital maintenance. Similar to, as horrible as it is, when a bridge collapses and people are killed, other bridges are repaired and maintained afterward.

It's just human nature to neglect routine maintenance. But when a threat comes along, it changes a person's perception. If you know someone is out to get you, you'll finally fix the lock on your front door—the same goes for cybersecurity. I want you to understand the risk and threats that are out there so you can respond appropriately and take the action needed to protect yourself, your employees, and your customers—it's a reality check.

Growth

Although growth is the goal of every business, it's also an impediment to trust. Because data is part of every aspect of your business, when you scale, every element—your employees, HR, finance, marketing, and leadership—has to scale the way they deal with privacy and security. Whether you grow organically or through mergers and acquisitions, you need to consider how it will impact your commitment to customer trust.

From the Ground Up

In an alternate universe, Zoom would have considered privacy and security as part of the DNA of their product during the designing phase—privacy and security controls should have been "baked in."

When building your product, you need to understand who your customers are and what type of sensitive information you'll be collecting from them—that is, *the fundamentals*.

Almost without exception, when a business creates a product, they focus on the customers and cost. Privacy and security need to be part of these initial discussions. Don't fall into the trap of having to redo something when you're almost ready to go to market. For example, in the EU, checkboxes on websites are required so people can opt in to receive emails. If the privacy conversation is skipped in the beginning, a website will have to go all the way back to the design phase. Save yourself time and aggravation by thinking ahead.

The variety of potential legal requirements (based on your business and the type of information involved) will shape your product. Things like where your privacy notice goes, what consent is needed, how to ask for that consent, vendor contracts, and how data is shared and secured all need to be considered. Basically, the best practice is to understand that data privacy and security usually go unseen but will make or break your business in the long run. So, treat it with the respect it deserves, from beginning to end.

M&A

Let's say you bought a mom-and-pop company for $5 million and you're in a rush to get the deal done. It's a brick-and-mortar operation, so you don't pay much attention to cybersecurity. They do have some data, but you're more interested in the customers coming into their retail location, so you just integrate their network into yours. Three months later, your company's network has a ransomware attack, and the demand is $10 million. Come to find out, there was an intrusion with the mom-and-pop shop before you purchased the business. When you integrated their network, the hacker gained access to yours as well. Now the $5 million company you just purchased is a $10 million-plus liability for your overall business.

Sound outrageous? This is exactly what happened when Marriott bought Starwood. Starwood had a breach that originated in its reservations system and had gone undetected for two years. When the two companies' systems were integrated, the hacker suddenly had access to both networks and thus to the personal data (including passport numbers) of customers from both companies, creating quite an event.

With company mergers and acquisitions, you need to exercise due diligence with privacy and security. A company could have their security locked down tight, but their privacy might be a little shady—maybe they aren't as transparent as they need to be, maybe they're nefarious. They could also be buying and selling data and improperly disclosing it. Get this figured out early because once the merger is made, their problems become *your* problems.

Check to see if the company you're interested in buying is complying with privacy laws; see how their privacy is managed; ask what kind of data is being collected; and understand how that data is being used. How the company is using data may not align with your trust-building strategy. In addition, think about how changes in technology might impact the deal. For example, not long ago Apple iOS and Google announced they would stop using third-party cookies. This meant businesses could no longer depend on them for ads. While we were writing this book, I (Jodi) fielded a call from a company whose business model was based on cookies and which didn't have a solid plan for the post-cookie world.

Company size is irrelevant here. I've seen small start-up teams with millions of records and large companies with surprisingly few. It's just about the data here: what do they have and how are they using it? Understand the flow of data. Look at the disclosures. See what privacy issues may exist. Many companies are bought just for their data; check for the use of opt-outs and disclosures. If their contract for the data doesn't mesh with your business goals, the merger will be for naught.

This also applies to companies that want to be acquired. Acquirers, now becoming more familiar with data security and privacy, will not buy a company if they are not compliant with certain privacy laws. Have your data ducks in a row if you are looking to sell.

One of the worst scenarios is that you'd buy a company, with their security and privacy looking good, only to find out their model is based on buying poor-quality data. You need to vet the source of the data as well. Lots of people just scrape data off the internet or buy it off a data broker—this is poor-quality data. There's no way to confirm its accuracy or validity. Good-quality data comes firsthand, like a company that has 100,000 people who've gone to their website and signed up for their newsletter—first-party contacts.

The source of the data will greatly affect your marketing efforts. If the data is poor-quality, purchased from some broker who got it from—*somewhere*, your open rates will be poor as well; the targets won't know who you are. (Plus, depending on the countries where they're located, your marketing to them might violate privacy laws.) They are also likely to be unqualified leads and not a part of your market. Compare that to a list of customers that know a brand and purposefully sought it out to give their information.

In addition to checking the quality of data, you'd be well advised to confirm the integrity of a company's security. Enter the "three-layer cake" (chocolate, of course) approach to cybersecurity due diligence. The first layer consists of three questions: Can you show me your network schematic, assets, and data flow? Do you have a written information security plan? Is there an organizational chart that shows how cybersecurity is reported to the C-suite and board of directors?

Typically, companies score zero out of three on this first layer. You'll need to ask these questions to both the C-suite executives and the IT department—the understanding of how a network operates is *vastly* different between the two. Get the right information from the right people. When I (Justin) speak at conferences, I always ask the audience what they would say if a buyer asked to have a third party rummage around your network during deal due diligence. They answer, "No!" I tell them that's the wrong answer. The right answer is, "Hell No!" This question brings into sharp focus the fact you will usually never get a full picture of network security during the due diligence process. That is why we have a second layer of our chocolate cake.

Layer two is all about the representations and warranties in your purchase agreement. You'll want confirmation that there have not been any data breaches, that all risk assessments have been disclosed, that the company is in compliance with its own privacy policies, that industry security standards are being met, and that "all our contracts are in full force and effect."

I recently added additional specific requirements like employees using MFA, password complexity, and the use of endpoint detection to the contracts I oversee. If the seller asks to strike these specific statements, it gives you the opportunity to ask why and learn more about omissions in their security hygiene. As your due diligence will likely never be complete, such "representations and warranties" are useful to fill the gaps relating to security. I also insist they survive the deal's closing by at least a year. Remember, a threat actor can get into your network and hang out for months before engaging in any mayhem.

Lastly, the third layer is to resist the urge to quickly integrate the company network you've just bought. Once you own it, have your security team thoroughly investigate and search the network for any intrusions, keeping it quarantined from your own network until that thorough check is complete. Too many companies simply integrate on day one and end up going through what Marriott did—learn from their mistake!

Mergers and acquisitions bring up unique risks. If you fail to obey the law, you risk a class action lawsuit or regulatory investigation. There was a company in Georgia that had superb security. But, when they acquired a company without the same levels of excellence, they went from having no problems to *many*. The acquired company was sanctioned and audited, leading to the purchasing company getting the same treatment—under which a problem was found in their company (the cherry on top). This company's reputation was damaged by an acquisition. All their hard work in building themselves up as a data-secure company was undercut in a moment—just as with Equifax and Marriott and their data-breach-conjuring acquisitions.

I once worked a ransomware case where an investor had poured $10 million into a company that had IT infrastructure circa 2005. Such old systems aren't patchable and thus are almost impossible to protect. They might as well have hung a sign on the front door that said: "*Please ransomware us!*" In the aftermath, when we tried to use our forensic software on their system to find how the inevitable had happened, it was so antiquated that our software couldn't talk to it, making our jobs extremely difficult.

It all comes back to companies not wanting to invest in maintenance—physical or digital. I get it. It's not nearly as exciting as closing a deal. The maintenance gets pushed back until there is an acquisition. If the right questions aren't asked in the due diligence phase, the onus is on the buyer. Why would a sophisticated investor buy a company that had technological infrastructure equivalent to a fifteen-year-old used car? Because they didn't think about cybersecurity, and even if they did, it was a fleeting thought. They didn't understand the basic concepts, let alone the potential risks. By now, I hope you're feeling clear about the latter. In the next chapter, we'll talk you through what you need to know to protect yourself.

PART FOUR: CYBERSECURITY

CHAPTER SUMMARY

Unlike dating, reimagining data as a medium for trust-building means protecting rather than sharing information. In both, understanding where you're vulnerable, what's at stake, and what your largest obstacles are is critical. Follow the three-layer chocolate cake approach to cybersecurity and privacy M&A due diligence so your deals deliver your just desserts.

CHAPTER 10
DEFENSE IN DEPTH

Once upon a time, there was a great king who ruled over a just and peaceful land. As the years went by and his knights went adventuring, the king collected a vast hoard of gold which he kept under the bed, and one day, thieves took it. The end.

Not very satisfying, is it?

And yet many CEOs, the royalty of our age, take about the same degree of care safeguarding their wealth which (in case I haven't been clear) is their data.

So let's try again.

Once upon a time, there was a great king who stored his treasure in a stone vault beneath his castle. His castle had tall walls surrounded by a moat accessible only by a drawbridge guarded by an army of well-trained and heavily armored guards. Thieves who were able to cross the drawbridge were stopped at the wall, and even those who got into the castle were thwarted by the guards or the vault. In cybersecurity, this practice of placing multiple barriers between valuables and malefactors is known as "Defense in Depth."

Some companies only have a drawbridge; when a hacker phishes an employee, the drawbridge is lowered and the hacker waltzes into the castle, free to roam. But if you have multiple layers of defense, the hacker might then run into a firewall before getting into the corporate network—perhaps this is the digital army in our dark-aged analogy. If the thief steals a key, he may find that it doesn't work on every door because a "least privileged policy" was put in place, limiting what employees have access to.

So many companies don't layer their defenses, making it incredibly easy for a hacker to do whatever he wants. If Target had this understanding years ago, their hackers would have only gotten as far as invoicing with the HVAC firm, and this pinpoints the data paradox. You want data to flow like water in your company, like underground plumbing, but you need to protect it at the same time, lest you have thieves using the sewers to navigate your fortress. Data allows you to learn so much about your customers, boosting your revenue. But, if it's mishandled or gets into the hands of the wrong people, data becomes a nightmare liability scenario similar to the one England's King Alfred the Great once faced.

King Alfred's daughter had been kidnapped and was being held for ransom by a band of particularly sophisticated Viking raiders. If he paid them what they asked, they would use the money to hire mercenaries to defeat him in battle and depose him. If he didn't, they would parade the naked princess through the country, using the fact of her capture and humiliation to undermine Alfred's authority and then depose him. (Or at least that's how it happened in a historical novel I read.) The point is that Alfred's wealth was what had allowed him to rise to greatness (he's the only English king to bear the sobriquet "the great") in his quest to unite England, but his gold (like your data) in the hands of enemies could have cost him (you) everything.

Let's take a more recent example. Once upon a time, Equifax wanted to become the largest data broker in all the land. In its hubris, Equifax developed one singular "foolproof" strategy: acquire as many companies as it could. Driven to achieve this goal as fast as possible, Equifax set about purchasing all manner of businesses and integrating their networks, completely disregarding security.[21]

A threat actor took notice and easily entered their network via what would have been an easily patched vulnerability, breaching their data hoard of 162 million records. In its great wisdom, Equifax first blamed the third-party software company through which the hacker had entered, but the truth of their own malfeasance was quickly discovered.

Once investigations were conducted, it was found that four of the senior officials had sold their Equifax stock before the breach was made public and that there had been over a month delay between the discovery and their response. That length of time may have been fine for a small private company but not for a large public company like Equifax. Think of how bad they looked once it was

---

[21] "An August 2016 report by the financial index provider MSCI Inc. assigned Equifax's data security efforts a rating of zero out of ten. The provider's April 2017 rating remained unchanged. Both reports concluded: 'Equifax's data security and privacy measures have proved insufficient in mitigating data breach events. The company's credit reporting business faces a high risk of data theft and associated reputational consequences...The company's data and privacy policies are limited in scope and Equifax shows no evidence of data breach plans or regular audits of its information security policies and systems.'" U.S. House of Representatives Committee on Oversight and Government Reform, *The Equifax Data Breach*, Majority Staff Report, 115th Congress, December 2018, 18, https://republicans-oversight.house.gov/wp-content/uploads/2018/12/Equifax-Report.pdf.

revealed that while millions of people conducted their daily lives not knowing their data had been exposed, executives were selling off their personal shares of the company. There is perhaps no better example of how to destroy customer trust!

BUILD YOUR FORTRESS

For a data company, Equifax was only focused on the business and growth side of its operations. They were all offense, no defense, raking in their digital gold and stashing it under the bed. Had they —or Zoom, or Marriott, or any of our previous example hacking victims—understood the basics of cybersecurity, all their issues could have been avoided.

In implementing a Defense in Depth strategy, I advise using the National Institute of Standards and Technology (NIST) 5-part cybersecurity framework it calls IPDRR for Identification, Protection, Detection, Response, and Resilience.

Identification

The first step of defense was easier (if bloodier) back in the good old days of Vikings and princesses. The thing you wanted to protect was that shiny gold stuff. Today, this step is necessary to identify your most important data and business processes.

In a law firm, this might well be the document management system. If it isn't available to the firm's lawyers, they can't do their work. Amazon, on the other hand, might be able to limp along for a while without its document management system. A denial of service attack that took down its website, however, would cripple it. Order processing is its critical business operation while a law firm would be only inconvenienced or embarrassed by its website going down.

A professional services company, on yet another hand, might survive either a disabled website or document management system but be utterly defeated if their Turbo-Tax-On-Steroids software locked them out.

For health care companies, patient records are going to be crucial while for others, their secure email system may be among the most critical of their business processes. For yet others, like Colonial Pipeline, it may be the software that controls machinery or some other aspect of manufacturing.

Having identified your critical systems and most sensitive data, you need to know where they are before you can protect them. Here, a regularly scheduled data inventory (see Chapter 5) is invaluable. We recommend once a year, but as a company grows and can afford automated tools, quarterly or even in real-time is preferable.

> Ask yourself: What are the vital business functions that, if disrupted, would cause the most catastrophic impact?

Protection

A good way to continue with these principles is to think about it as protecting your children in your home. You've already identified who you are protecting, and now you need to think about *how* to protect them. Ask yourself how they might be harmed in your home. If an intruder comes through the window, put locks on the windows. You get the picture.

In business, it works the same way. Having identified what's more important, ask yourself how someone might access or damage it. Consider each of the four main threat vectors we discussed in Chapter 8: internet, wireless, supply chain, and insider threat.

Since the biggest threat* comes from your own employees clicking on attachments and going to websites they shouldn't, let's say they represent the front door. Is it standing wide open? Train your workers to recognize phishing attempts and social engineering. Education is step one and goes a long way in fighting against threat actors.

PART FOUR: CYBERSECURITY

> *Even if there's just one mistake or one phished employee, a hacker can get in, and it's rarely just one. Over a third of employees[22] click on phishing emails. See the section in Chapter 8 on Insider Threats.

The more steps you can add, the more layers you have, and the more you build your Defense in Depth. More is definitely better here. Segment your network, enact least privilege access, and use multi-factor authentication—it's your house, protect it like you ought to.

Implementing a firewall is another excellent way to protect these access points. Firewalls detect what is coming into a network and block whatever isn't pre-approved. Endpoint detection also helps you identify unusual traffic coming into your network from the outside. Think of it as an additional alarm system besides your firewall.

---

[22] Perry Carpenter, "KnowBe4's 2021 Phishing by Industry Benchmarking Report Reveals that 31.4% of Untrained End Users Will Fail a Phishing Test," Security Awareness Training Blog, *KnowBe4*, July 8, 2021, https://blog.knowbe4.com/knowbe4s-2021-phishing-by-industry-benchmarking-report-reveals-that-31.4-of-untrained-end-users-will-fail-a-phishing-test.

Cybersecurity is centered around protecting the same three priorities: confidentiality, integrity, and access (CIA), discussed in Chapter 5. Identity Access Management (IAM), as you may remember from Chapter 6, is a combination of technologies* and policies that determine who can access the computer network and how broad that access is. The latter is what businesses forget about far too often. Companies may have authentication practices in place for data access, but once inside, employees can roam wherever they please. Segregating your network and isolating the particularly sensitive parts creates additional layers of security.

> *Technology can help automate the Identity Access Management policies your organization develops. Okta, Onelogin, Cyberark, and Microsoft all sell excellent IAM tools, and third-party managed service providers often include them as part of their overall offering.

Far too many hacks occur because once the hacker gets into the network (think Target again), they gain complete access to the entire network. That is why the administrative assistant to the CEO or CFO is such an inviting hacking target. Carefully consider what access the CEO's administrative assistant needs to perform his job function. If he has complete access to the CEO's email inbox, hacking him is equivalent to hacking the CEO's inbox and likely a much easier target.

> It's critical for every organization to develop and enforce an IAM policy. The best-laid security plans fail due to a lack of good Identity Access Management. It is a key component of the defense-in-depth concept since no company can completely defend itself against people clicking on emails!

Detection

Even the most secure system can be breached. To return to our home defense analogy, if clever thieves laser a hole in a window, reach in, and unlock it, it's the difference between having an alarmed window that wakes you up when they ease it open and sleeping through the night. The story of the Target hack illustrates what can go wrong when there's a failure in the detection portion of the NIST framework. The company had endpoint detection software* that alerted people to an intrusion, but they ignored its warnings.

It's critical to recognize the vulnerability of any place where your company's data comes into direct contact with the larger internet. Get the right people to help you tailor a solution to your organization.

> *Network computers and other devices directly connected to the internet are prime entry points for malware and ransomware. Endpoint detection tools are designed to spot suspicious activity on these endpoint devices and sound the alarm.
>
> Companies that offer popular endpoint detection tools are Crowdstrike, Sentinel One, Microsoft, and Sophos. If you are a smaller company and hire a managed services provider, they may provide endpoint detection as part of their suite of services. For enterprise-level businesses, Crowdstrike's tools are very good but cost-prohibitive for small or lower middle-market organizations.

Response

So the thieves are in your home, and you know it. Now what? Shotgun? Saferoom? A lack of preparation compounds the damage of any attack. Ransomware hackers count on causing time-stressed, unorganized panic. The bigger the problem they create, the more they can collect in ransom. Companies that have a response plan in place ahead of time save, on average, $300,000–$500,000 in an attack.*[23]

An incident response plan* helps you classify events and respond appropriately—no need to wake up the CEO if a laptop is stolen, but sound the alarm if your network gets encrypted.

> *To get your company started, check out the free downloadable incident response plan template on the Data Reimagined website DataReimaginedBook.com

A good incident response plan categorizes incidents based on severity, from accidentally emailing the wrong person sensitive information and breaching privacy to a full-blown ransomware event—obviously, this assumes a company has completed its data inventories and threat analyses. It tells employees who to report to when an incident occurs in an effort to catch the problem early and mitigate the consequences.

It's just like a school fire drill or the emergency demonstrations

---

[23] "How to Limit the Cost of Data Breaches," Compliancy Group, October 8, 2019, https://compliancy-group.com/how-to-limit-cost-of-data-breach.

before each plane flight. The incident here is the emergency, and the response plan is the drill; like the old adage says: *failing to plan is planning to fail.* Companies need to understand the risk, see the potential dangers, think through the response plan, document that plan, and keep it current. Ideally, they'd conduct surprise simulations and scenario drills to give their people practice using the response plan.

Such advance planning streamlines the process and avoids any confusion on who does what since people already have defined roles and responsibilities. The CEO may be the decision-maker, but PR needs to handle the press release, and IT will be the ones working with the forensics team. With a hierarchy established, reports to the decision-makers will be timely and full of relevant information.

Get as detailed as you can in your response plan as far as levels of data, who to contact, and roles and responsibilities go, but I advise against plans that are purely in an "if A then B" type format ("A" rarely happens as written out). Cybersecurity incidents come in all sorts of shapes and sizes, and you need an adaptable, flexible plan to respond. I've never seen a company, mid-attack, reading a thirty-page document.

We'll discuss assembling a cybersecurity team in the next section, but here it's important to mention including the contact information of all involved third parties (e.g., outside counsel, forensic firm, ransom negotiator, federal and local law enforcement, insurance carrier, etc.) in your incident response plan. Gather any information you'll need ahead of time to avoid scrambling and being overwhelmed by all you'll need to do. Print it out. If it's just another file on your system, the hack (the reason you need it) may keep you from being able to access it.

No company is too small to have a plan. At a minimum, know who your team is, how to contact them, and what each person's role and responsibility is.

Resilience

Resilience is the strength built through adversity. It's how to make sure no great crisis gets wasted. Build a strong post-mortem into your security framework so that your company learns from its mistakes. Like an after-action report in the military, a post-mortem brings together the team that responded and reviews their effectiveness —identifying what worked, what didn't work, and what might be done better the next time.

Sometimes, if the gaps are huge, an independent risk assessment by a third party can surface additional gaps. This can lead to a prioritized remediation plan with implementation milestones to help improve cyber hygiene. In my (Justin's) experience, such large gaps are most frequently caused by a failure to separate security from IT (as recommended in Chapter 2). Too often, it's the expense of bringing in these outside consultants that finally convinces the C-suite to invest in hiring or outsourcing its cybersecurity.

Being resilient also requires you to evaluate how you will continue to evolve your security posture as new technology (and thus new threats) emerge. What's the plan if enemies show up at your castle walls with a heavily armed drone?

> **A strong defensive framework will help you identify what to protect, how to protect it, detect intruders, respond to breaches, and create resilience.**

MARSHALL YOUR FORCES

The aftermath of a data breach is not the right time to start building your defense team. You need a team of people in place long before a threat actor even thinks about your systems. You can't afford to be spending time shopping around for vendors that your insurance carrier has pre-approved while all your systems are encrypted.

You could spend two to three days trying to find everyone you need; meanwhile, the hacker is doing who-knows-what to your network. Don't get me wrong, businesses do it the wrong way all the time; it's basically the bread and butter of my business—they're in a scramble and I'm the guy they've heard about, but it's far from ideal.

In a perfect world, all businesses would have their own privacy and security attorneys, but you should at least have already vetted some third-party legal specialist whose cost will be covered by insurance. Certainly, your internal counsel should be educated in this space and will likely be the first person on your incident response plan's notification list. Typically, your legal department will be calling the shots, and incident responders will be engaged through outside legal counsel for the sake of attorney-client privilege.

Cybersecurity Insurance

Cybersecurity insurance is meant to help you recover from the damages of a cybersecurity incident and their associated costs, which can include legal costs, forensic teams, communication firms, ransom negotiators*, breach teams, rebuilding your systems, regulator penalties, and lawsuits.

> *Ransomware negotiators make contact with the threat actor and facilitate the payment in Bitcoin.

Since early 2020, insurance companies have been battered with ransomware attack claims, and in one case an insurer itself was hacked and paid $40 million to get their own network back.[24] As a result, many insurance providers have pulled out of the cybersecurity market, and those that remain are raising their premiums while lowering their coverages. On top of this, unlike more mature lines of insurance, cybersecurity insurance is relatively new; the policies from carrier to carrier aren't uniform. I still recommend it if for no other reason than it forces companies to maintain a certain level of security as a stipulation of the policy.

Find a broker who understands the product to make sure that your policy provides the coverage that your business needs.* Not all policies are equally good, and the one you're considering might not cover your specific circumstances. This is a point where I recommend calling a professional—you need someone who knows what they're doing and is intimately acquainted with the risks involved. So many

---

[24] Kartikay Mehrotra and William Turton, "CNA Paid $40 Million in Ransom After March Cyber Attack," *Insurance Journal*, May 21, 2021, https://www.insurancejournal.com/news/national/2021/05/21/615373.htm.

cyber insurance brokers out there don't understand what they're selling. This would also be a good way to test your cybersecurity lawyer; see if they can help find a broker and assess what coverage makes sense for your business.

> *For a free list of the right questions to ask about a cybersecurity insurance policy, visit DataReimaginedBook.com

Privacy

Some companies start off with an understanding of privacy law and give that responsibility to their general counsel. Later they'll see the need for a specialized privacy attorney and even choose to make that person their key privacy officer.

Other companies begin from more of a security standpoint, and hand privacy matters off to their security team, associating data breaches with privacy. Companies will quickly realize the stark differences between privacy and security as the security team now must ask the marketing department what they're *doing* with the data instead of just protecting it. I've spoken to CMOs that wound up taking over privacy after discovering how much was involved in having a privacy notice or cookie banner.

For massive companies, privacy might fit into their legal department, compliance group, or risk group. They'll also have privacy "champions" placed throughout the company—people who herald privacy's importance. The marketing department may have a privacy guru, or HR might have their "go-to" person.

Smaller companies may not have the resources for so many positions, but the idea remains the same. An external part-time privacy expert can be hired, or an internal position can be given the responsibility.

> Justin: You mean someone like you?
>
> Jodi: Yup! I serve as the Fractional Privacy Officer for many companies.
>
> Justin: I know you do, and you're great at it!
>
> Jodi: Awww, that's so nice of you.
>
> Justin: Yeah, I'm a sweetheart.
>
> Jodi: (rolls eyes)

It depends on the company where you choose to place this responsibility, but typically you want the person to have the influence to create a privacy initiative effectively.

Oftentimes, the head of privacy will be your point of contact person who hires consultants to come in and help with the tactical and technical work. Companies nowadays are using Virtual Chief Information Security Officers (vCISOs). Even though they're virtual, vCISOs need to be held accountable and coordinate with someone in the company.

Privacy isn't just that one person's job; it's the entire company's. Your head of privacy needs to rally all parts of your business to ensure privacy compliance. Regardless of the direction, you need this person to have some executive-level sponsorship, someone to advocate the importance of their message.

Leadership

Both privacy and security must be led from the top and by example. The bottom line is that whatever way C-suite treats security and privacy, the company follows. We have a friend who's a CEO. When we spoke to him about these principles, he said, "Oh, I'm the CEO. Those rules don't apply to me." Obviously, there was a disconnect, and his company suffers for it. "Do as I say, not as I do" has never worked, especially with data privacy and security. If you really want to change the culture in your company, you need to step up, set the tone with your actions, and hold your employees accountable.

Privacy and security need to become part of your company's culture, and it starts at the top. If you're the leader of your company, you set its trajectory on all fronts. Show your employees the importance of data privacy and security in your actions. I've seen CEOs wake up to these concepts, make radical moves, and completely change the culture of their business.

One of the easiest ways you can start the change is by creating company videos on the importance of data security and privacy, connecting it to your core principles. You can give your privacy and security teams an adequate budget and allow them room to speak in meetings. Talk to them directly and consult them on risks—treat them like the core business function they are.

Mercenaries

Remember King Alfred and his concern that any gold he paid to ransom his daughter would be used to pay for soldiers to fight against him? Swords for hire have been a staple of warfare for centuries, but the term "mercenary" has gradually picked up a negative connotation it doesn't deserve. If a company is unable to provide its own security team, then outsourcing is a great answer. A managed service provider (MSP) can provide both functions and report to you just like an in-house team would. Of course, outsourcing comes with its own risks and creates another way of being hacked. Since smaller companies really don't have any other choice than hiring an MSP, our advice is to thoroughly vet your vendor.

Finding a good MSP isn't easy; my foremost recommendation is to get a good advisor. There's not a master list of questions to ask when doing your due diligence here—everything is dependent on your business and product.

WASH YOUR HANDS

In the world of data privacy and security, the word "hygiene" is used to describe those few basic practices that can improve your immunity to the trust-destroying plagues that accompany our massively interconnected technology. Think of this list as the data equivalent of the advice to lock your car and take your keys—basic safety precautions that will help you safeguard your most important digital assets.

- Use a phrase (not a single word) for all your passwords that's at least nine characters long and includes at least one capital letter and one special character.
- Don't use the same password on multiple sites.
- Use multifactor authentication.

- Don't write passwords down. We recommend a password manager like Google's, Keychain, or LastPass.

- Perform a basic data inventory and keep it current.

- Have and regularly update your website's privacy notice and post the date of its latest iteration.

Designate responsibility for data privacy to a specific person and make another person responsible for cybersecurity.

Access the internet securely—don't use public Wi-Fi and use strong password protections on your routers.

- Use the latest operating system for your phone.

- Keep all your software up to date.

- Never click on emailed or texted links.

- Assume any request for money or account access is fraudulent.

- Use antivirus software.

Configure the privacy and security settings of websites and software tools that house company data.

CHAPTER SUMMARY

Implementing a Defense in Depth strategy of layered protections is the best way to defend your company and your customers' data from hackers. Although anything that creates impediments to unauthorized access introduces a little friction in authorized use, it's not that different from taking the additional second to buckle your seat belt. Failure to put such precautions in place will only add insult to injury in the eyes of your customers should hackers gain access to their data through your system.

Identify your most business-critical data and processes, consider the threat vectors to which they're most vulnerable, and implement measures to protect their confidentiality, integrity, and accessibility. Think of these measures as deterrents more than preventatives. Make sure your people—both potential insider threats and leadership—are trained in best practices and walking their talk. Have both properly trained people and the right software to detect a breach if (when) one occurs. Provide a clear incident response plan that details who to contact and what to do in the immediate aftermath, and after the crisis has passed, conduct a post-mortem to increase resilience against future attacks.

CONCLUSION

Some two hundred years ago, the Industrial Revolution changed the way people lived, worked, and interacted. For the first time in human history, more people lived in cities than small towns and regularly encountered more strangers than friends. These tectonic shifts gave rise (among other things) to new forms of crime. With goods stored in warehouses and money stockpiled in banks, there were new targets of opportunity and a new need for professional police and trade regulations.

While it might be pushing things to claim that banks create bank robbers or that overseas trade spawned pirates, since the days when people used to bite a coin to literally test its metal, that which is valuable has been vulnerable to theft or degradation and trust has been vital to commerce.

Our goal in writing this book has been to help readers recognize the same is true in the wake of the Fourth Industrial Revolution. As data has become the currency of the Information Age, new industries and new forms of crime have cashed in on its rising value. To meet the challenge of the ghost of data future, business leaders need to change how they think about it. They need to reimagine data.

Having worked in the vanguard of the data revolution, we're convinced that its second wave will follow the example of other new technologies. In the same way that speed limits followed the building of highways and brand loyalty became a stand-in for a personal friendship with a local merchant, government regulation and new forms of trust-building are the future.

In *Reimagining Data*, we've argued that the gold rush mentality of collecting as much customer data as you can grab doesn't translate into having a relationship with customers any more than stalking leads to love. But data collection that is respectful and consensual can build trust, which is an even more valuable (and increasingly rare) commodity.

But it isn't easy.

Many of the trust-building practices we recommend come at a cost in time, money, and efficiency. Trust builds slowly and leaves fast. It requires the implementation and maintenance of practices that introduce complications and delays and are more focused on prevention than productivity. Happily, these very difficulties increase its value on three fronts: It becomes a differentiator—something

CONCLUSION

you can offer that your competitors don't. It increases the amount of data your customers may willingly share with you, making every advertising dollar more targeted and effective. It puts you ahead of the future costs of instituting sound privacy and security practices once legislation and early adopters make them mandatory.

Reimagine data privacy and security as a way of demonstrating your respect for your customers. Build privacy and security into your business processes, from design to products to culture. Articulate and document this new mindset in policies and procedures that stipulate collection of only data for which you have a business use, which account for its privacy and security, and which comply with the wide range of laws that govern customers' rights, breach notification, and the purchasing of data.

Inventory where your data is kept and how its confidentiality, integrity, and availability are secured and access-regulated, accounting for the relative advantages of cloud or network storage. And keep this documentation up to date.

Communicate your commitment to building trust through data privacy and security to your customers. Allow them to dictate their preferences and ensure that their wishes and your policies are followed by your employees, contract workers, and third-party vendors abide by it.

But of course, these aren't the only people with an interest in your customers' data. Hackers with diverse motivations can employ an equally diverse range of hacks to gain or block access to your data, and it's your responsibility to protect your customers and yourself from them.

Given their audacity, creativity, and tenacity, defense against hackers requires layers of protection. Identify what is most vital to protect, where it's most vulnerable, and implement measures to protect its confidentiality, integrity, and accessibility. A single breach can destroy the trust you've taken such pains to build, and a ransomware event can empty the coffers.

You won't be able to do all this at the same time. You'll need to create a plan and start with simple steps, or else you'll be overwhelmed. Designate two internal "champions," one charged with safeguarding your company's cybersecurity and the other with advocating for data privacy. From there you can bring in outside help and build up a team piece by piece. Changing your company's culture to be more privacy- and security-minded is no small task. But we believe that any sized company can create a privacy and security program that works for them, so go in with the belief that you *can* change your company. It starts with changing how you think.

The protagonist of *A Christmas Carol* woke up from his nightmare of ghosts and Christmases and ran right out to buy a goose. We

CONCLUSION

hope you never have to experience any of the many nightmare scenarios we've described. We hope you join us in imagining a future in which data is no longer scraped and hoarded, Scrooge-like, but used with respect to build trust, deepen relationships, protect everyone's privacy, and enhance our collective security.

CPSIA information can be obtained
at www.ICGtesting.com
Printed in the USA
LVHW102240151122
733215LV00019B/526/J